Human Factors in Healthcare

A Field Guide to Continuous Improvement

Synthesis Lectures on Assistive, Rehabilitative, and Health-Preserving Technologies

Editor

Ron Baecker, *University of Toronto*

Advances in medicine allow us to live longer, despite the assaults on our bodies from war, environmental damage, and natural disasters. The result is that many of us survive for years or decades with increasing difficulties in tasks such as seeing, hearing, moving, planning, remembering, and communicating.

This series provides current state-of-the-art overviews of key topics in the burgeoning field of assistive technologies. We take a broad view of this field, giving attention not only to prosthetics that compensate for impaired capabilities, but to methods for rehabilitating or restoring function, as well as protective interventions that enable individuals to be healthy for longer periods of time throughout the lifespan. Our emphasis is in the role of information and communications technologies in prosthetics, rehabilitation, and disease prevention.

Human Factors in Healthcare: A Field Guide to Continuous Improvement
Avi Parush, Debi Parush, and Roy Ilan
2017

Assistive Technology Design for Intelligence Augmentation
Stefan Carmien
2016

Body Tracking in Healthcard
Kenton O'Hara, Cecily Morrison, Abigail Sellen, Nadia Bianchi-Berthouze, and Cathy Craig
2016

Clear Speech: Technologies that Enable the Expression and Reception of Language No Access
Frank Rudzicz
2016

Designed Technologies for Healthy Aging
Claudia B. Rebola
2015

Fieldwork for Healthcare: Guidance for Investigating Human Factors in Computing Systems
Dominic Furniss, Rebecca Randell, Aisling Ann O'Kane, Svetlena Taneva, Helena Mentis, and Ann Blandford
2014

Fieldwork for Healthcare: Case Studies Investigating Human Factors in Computing Systems
Dominic Furniss, Aisling Ann O'Kane, Rebecca Randell, Svetlena Taneva, Helena Mentis, and Ann Blandford
2014

Interactive Technologies for Autism
Julie A. Kientz, Matthew S. Goodwin, Gillian R. Hayes, and Gregory D. Abowd
2013

Patient-Centered Design of Cognitive Assistive Technology for Traumatic Brain Injury Telerehabilitation
Elliot Cole
2013

Zero Effort Technologies: Considerations, Challenges, and Use in Health, Wellness, and Rehabilitation
Alex Mihailidis, Jennifer Boger, Jesse Hoey, and Tizneem Jiancaro
2011

Design and the Digital Divide: Insights from 40 Years in Computer Support for Older and Disabled People
Alan F. Newell
2011

Human Factors in Healthcare: A Field Guide to Continuous Improvement
Avi Parush, Debi Parush, and Roy Ilan

ISBN: 978-3-031-00474-2   print
ISBN: 978-3-031-01602-8   ebook

DOI 10.1007/978-3-031-01602-8

A Publication in the Springer series
SYNTHESIS LECTURES ON ASSISTIVE, REHABILITATIVE, AND HEALTH-PRESERVING TECHNOLOGIES #11

Series Editors: Ronald M. Baecker, University of Toronto

Series ISSN: 2162-7258 Print   2162-7266 Electronic

Human Factors in Healthcare

A Field Guide to Continuous Improvement

Avi Parush
Israel Institute of Technology
Debi Parush
UPT: Usability Publications and Training
Roy Ilan
Queen's University, Canada

SYNTHESIS LECTURES ON ASSISTIVE, REHABILITATIVE, AND HEALTH-PRESERVING TECHNOLOGIES #11

ABSTRACT

Have you ever experienced the burden of an adverse event or a near-miss in healthcare and wished there was a way to mitigate it? This book walks you through a classic adverse event as a case study and shows you how.

It is a practical guide to continuously improving your healthcare environment, processes, tools, and ultimate outcomes, through the discipline of human factors. Using this book, you as a healthcare professional can improve patient safety and quality of care.

Adverse events are a major concern in healthcare today. As the complexity of healthcare increases—with technological advances and information overload—the field of human factors offers practical approaches to understand the situation, mitigate risk, and improve outcomes.

The first part of this book presents a human factors conceptual framework, and the second part offers a systematic, pragmatic approach. Both the framework and the approach are employed to analyze and understand healthcare situations, both proactively—for constant improvement—and reactively—learning from adverse events.

This book guides healthcare professionals through the process of mapping the environmental and human factors; assessing them in relation to the tasks each person performs; recognizing how gaps in the fit between human capabilities and the demands of the task in the environment have a ripple effect that increases risk; and drawing conclusions about what types of changes facilitate improvement and mitigate risk, thereby contributing to improved healthcare outcomes.

KEYWORDS

human factors, ergonomics, healthcare, patient safety, quality improvement, adverse events, human error, interventions, mitigations

Contents

1 Background to Human Factors in Healthcare **1**
  1.1 Healthcare Scenarios 1
      1.1.1 A Simple Retrospective Case—Nurse in the Intensive Care Unit ... 1
      1.1.2 A More Complex Retrospective Case 2
      1.1.3 A Failed Case—in Retrospect 2
      1.1.4 A Proactive Case 3
      1.1.5 The Factors behind the Cases 3
  1.2 We Need a Human Factors Perspective 3
      1.2.1 Introducing Technology to Healthcare Adds Complexity 4
      1.2.2 Human Factors can Contribute to Patient Safety and
            Medical Error Prevention 4
  1.3 About the Book 4
      1.3.1 Target Audience 5
      1.3.2 A Different Kind of a Human Factors Book 5
      1.3.3 Book Organization 6

Part I: A Conceptual Framework **7**

2 About Human Factors Frameworks **9**
  2.1 Why Have a Framework? 9
  2.2 Frameworks of Human Factors in Healthcare—A Review 9
      2.2.1 The SEIPS Model—Systems Engineering Initiative for
            Patient Safety 9
      2.2.2 Human Factors Engineering Paradigm 11
      2.2.3 A Human Factors Framework for Analyzing Risk and
            Safety in Clinical Medicine 11
      2.2.4 The FAA's Human Factors Analysis and Classification System
            (HFACS) 11
      2.2.5 The WHO Human Factors Framework 12
      2.2.6 The Food and Drug Administration Human Factors Framework .. 12
      2.2.7 A Summary Comparative Table 13
  2.3 Critique and Summary 13

**3 HF—MARC: Using the Human Factors Conceptual Framework
to Map-Assess-Recognize-Conclude** **15**
  3.1 Overview 15
  3.2 HF Conceptual Pyramid: Rationale 16
  3.3 Methodological Process: Rationale 18
  3.4 A Walk through the HF-MARC Framework 20
    3.4.1 Map the Foundation Tier: People, Tasks, and Environments 20
    3.4.2 The Moderating Tier: Assess the Fit and Recognize
      Emergent Factors 23
    3.4.3 The Top Tier: Conclude about Performance and Outcomes 25
    3.4.4 Interventions and Mitigations 26

Part II: The Human Factors Field Guide **29**

4 Overview: A Human Factors Approach to Continuous Improvement **31**
  4.1 Recap of the framework, methodology, and a roadmap 31
  4.2 The Sample Scenario 33

5 Start the Process **35**
  5.1 Determine Objectives and Scope of the Analysis 35
  5.2 Use Checklists for the Mapping and Analyses 36
    5.2.1 The Short Mapping Checklist 37
    5.2.2 The Detailed Checklist 38
  5.3 Choose the Checklist for the Analysis 38

6 Map the Context **41**
  6.1 People 41
    6.1.1 Which Populations and Segments are Relevant to the Analysis ... 44
    6.1.2 Persona 44
  6.2 Missions, Goals, and Tasks 46
    6.2.1 Task Analysis: Map the Inter-relations between the Tasks 48
    6.2.2 Task Analysis Implications to Severity Ratings 51
    6.2.3 Task Flow Mapping 51
  6.3 Physical Environment 53
    6.3.1 Space and Layout 53
    6.3.2 Artifact and Device Locations and Access 56
    6.3.3 Ambient Conditions 58
    6.3.4 Human-Machine Interface 60

6.3.5   Mapping Human-Machine Interface Using the
Detailed Checklist 61

6.4   Human environment 64

6.5   Interim Brief #1 68

7   **Assess Fit** **73**

7.1   A Brief Overview of Some Relevant Human Capabilities and
Limitations 74

7.2   Assessment Criteria 82

7.3   People and Tasks 84

    7.3.1   The People Perspective 84

    7.3.2   The Task Perspective 86

7.4   Physical Environment 87

    7.4.1   Space and Layout 88

    7.4.2   Location of Artifacts and Devices 89

    7.4.3   Ambient Conditions 90

7.5   Device Usability 91

7.6   Human Environment 94

    7.6.1   Groups and Teams 95

    7.6.2   Organizations, Climates, and Cultures 97

    7.6.3   Rules, Regulations, and Policies 98

8   **Interim Findings: Problems of Fit** **99**

8.1   List of the Fit Problems by Factors 99

8.2   Problem Severity 102

9   **Recognize Emergent Factors** **105**

9.1   Developments that May Influence the Performance and Outcomes 105

9.2   Emergent Environmental Factors 106

    9.2.1   Workload 106

    9.2.2   Distractions and Interruptions 106

9.3   Emergent Human Factors 107

    9.3.1   Mental and Physical Workload 107

    9.3.2   Discomfort 109

    9.3.3   Fatigue and Loss of Vigilance 109

    9.3.4   Stress 110

    9.3.5   Stress, Fatigue, and Loss of Vigilance—A Synthesis 111

9.4   Summary of Emergent Factors 112

    9.5   Interim Brief #2 115

10 **Conclude: Performance and Outcomes** **117**
    10.1 The Most Influential Factors 117
        10.1.1 Scope According to Validity and Relevance 118
        10.1.2 Identify the Most Influential Factors 118
    10.2 Performance and Outcomes: How are they different? 121
        10.2.1 Performance: Effectiveness, Human Error, and Efficiency 122
        10.2.2 Outcomes: Factual, Likely, and Desired 126

11 **Interventions and Mitigations** **131**
    11.1 Granularity of the recommendations 131
    11.2 Intervention and Mitigation Strategic Goals 132
    11.3 Prioritize the Interventions Using a Tradeoff Analysis 136
    11.4 Final Brief #3 138

12 **This is Not the End** **141**

    Appendix A: The Detailed Checklist **143**

    Appendix B: Fully Analyzed Sepsis Management Scenario **155**

    References and Resources **187**

    Author Biographies **201**

CHAPTER 1

Background to Human Factors in Healthcare

You may wonder how to fit the discipline of human factors into the already overloaded domain of healthcare. Healthcare professionals are dealing with a multitude of technological advances and information overload. The amount and rate of change challenge their ability to use their knowledge for saving lives and doing no harm. While these challenges are true for most modern domains, the complexity and criticality of healthcare intensify these challenges. By analyzing the human factors involved, it is possible to improve the tools, processes, and care environment, ultimately improving healthcare outcomes.

The scenarios below can help you discover where human factors appear. There are many statistics on failures in healthcare. Many of these failures began with problems in human factors. Adopting a human factors perspective provides a huge opportunity to improve patient health and safety. Continue reading to learn how.

This book addresses you as a player in the healthcare system who can contribute through greater awareness, a systematic approach, and a powerful human factors analysis toolkit. It can also be used by professionals interested in improving other complex, critical services using a human factors approach. While we wrote the book with the intention of giving you all the information you need to use the book as an analysis tool, we offer references and additional sources for those who would like to learn more.

1.1 HEALTHCARE SCENARIOS

The following scenarios describe a variety of retrospective and proactive cases. In the retrospective cases, we already know the outcomes, both the successes and failures. In the proactive case, we describe an example that involves planning for the future in which the outcome has not yet occurred. These examples portray possible contexts for the human factors approach in healthcare.

1.1.1 A SIMPLE RETROSPECTIVE CASE—NURSE IN THE INTENSIVE CARE UNIT

Consider a simple case. A nurse in an Intensive Care Unit (ICU) walked over to check the vital signs of a patient. The nurse succeeded in this task: she noticed all the parameters on the monitor, understood the status of the patient, and knew what to do next.

Can you think of some reasons for her success?

You may suggest that the nurse was experienced and it was not the first time she checked vital signs. Maybe it was quiet in the ICU and there were no distractions or interruptions so she managed to focus on the information and use it adequately. Maybe she was also fresh and alert after a long rest before coming to work her shift. Probably a clear visual presentation of the information on the screen of the monitor helped. Can you think of any other reasons?

1.1.2   A MORE COMPLEX RETROSPECTIVE CASE

Let's look at a more complex case. The surgery team in the operating room, in the midst of open-heart surgery, was preparing to arrest the beating heart of the patient before they initiated the bypass to the pump. The anesthetist was ready to induce cardioplegea. The perfusionist was ready on the pump. The surgeon was performing the last steps before this critical phase in the open-heart surgery. Everything looked OK and ready.

What led to the team's effectiveness and the successful progress of this operation?

We like to start by suggesting that the members of the team were experienced, and were used to working with each other. They may have communicated so far in an effective manner and no one on the team was missing any information. There were not any interruptions or distractions. The information was very clear on the monitors and all the required tools were available.

1.1.3   A FAILED CASE—IN RETROSPECT

How about looking at a case where things did not end up well? Paramedics brought a man with multiple gunshot wounds into the emergency department (E.D.). The emergency physician on duty started immediate resuscitation with the help of a nurse, while they called for additional assistance. The patient's status deteriorated: his blood pressure dropped rapidly and oxygen level was low. Additional assistance came too late and the team had to pronounce the patient dead after a long resuscitation attempt.

Can you think of some reasons for this failure?

One immediate possibility you could suggest is the clinical status of the patient and it cannot be categorized as an adverse event. But there could be other factors that play a role here. Maybe the ED team did not get all the necessary information from the paramedics. Maybe the ED physician called for the additional help too late. Maybe because of the commotion that the patient's family created, the nurse did not follow the vital signs closely. Maybe the physician had not slept for the past 36 h because he took on an additional shift due to lack of personnel. Can you think of any other reasons?

1.1.4   A PROACTIVE CASE

We took a retrospective look at the events and situations described above, including successes and a failure. Here is one last case, for now, that is *proactive*: a hospital plans to implement a new information system.

What factors should you consider when implementing the system to ensure successful acceptance and effective use?

You may want to consider who will be using the system, what task they will be performing with it, and their previous experience with that task.

1.1.5   THE FACTORS BEHIND THE CASES

When considering the reasons for success or failure in a retrospective case, you can probably think of many factors. In anticipation of contributing factors in proactive cases, you probably have important intuitions as well. And when you have exhausted all the reasons and factors you can think of, do you suspect there may be other reasons you have not thought about?

Probably....

Identifying the reasons for the outcomes is critically important because the better we understand the factors that influence a case, the more we will be able to replicate the successes, on the one hand, and prevent or mitigate the failures, on the other.

In the above retrospective and proactive cases, we probably manage to intuitively identify some or even most of the factors that are relevant and influential. At the same time, we may miss some. This exposes a need for a systematic approach, a framework, to help us discover all the influential factors, and then use that discovery to ensure success while preventing or mitigating failures.

1.2   WE NEED A HUMAN FACTORS PERSPECTIVE

The problem of overlooking important factors that could prevent adverse events may be widespread. Preventable adverse events, in which patients are harmed, occur too frequently. Internationally, patient safety and medical errors are increasingly being recognized as critical problems. According to the frequently cited report "To err is human: Building a safer health system" (Kohn et.al. 2000), between 44,000 and 98,000 Americans die each year as a result of medical errors. "The Canadian Adverse Events Study: The incidence of adverse events among hospital patients in Canada" (Baker, et al., 2004) reported that 7.5% of hospital admissions result in adverse events which harm patients. Of these adverse events, 37% were potentially preventable. A little more than ten years later, Baker and Black (2015) published a follow-up report titled: "Beyond the quick fix: Strategies for improving patient safety", in which they report that despite efforts to improve, many Canadian healthcare organizations still struggle to prevent patients and their families from experiencing harm. A more recent article suggested that prior reports on adverse events was underestimated and that medical errors was the third leading cause of death in the U.S. (Makary and Daniel, 2016).

Human factors can play a key role in mitigating risk and improving the situation.

1.2.1   INTRODUCING TECHNOLOGY TO HEALTHCARE ADDS COMPLEXITY

Healthcare organizations are often considered as complex systems (e.g., Lipsitz, 2012; McDaniel et al., 2013). Introducing technology to healthcare adds to the complexity and poses challenges that may result in medical errors (for example, Ash et al., 2004a; Kohn et al., 2000). Kohn et al. (2000) report that about 71% of the preventable adverse events occur in the operating room. Several contributing authors in "Human error in medicine" (Bogner, 1994) suggest that many errors are associated with poor design of the medical devices and equipment used by the medical professionals in general, and in the operating room in particular. Thus, human factors are a key aspect in the use of technology in healthcare.

1.2.2   HUMAN FACTORS CAN CONTRIBUTE TO PATIENT SAFETY AND MEDICAL ERROR PREVENTION

For quite a few years there has been an increasing awareness and practice of cognitive and human factors engineering in healthcare (for example, Carayon, 2016; Carayon et al., 2014a; Gosbee, 2002; Gurses, and Pronovost, 2011; Wachter, 2012; Weinger and Slagle, 2002; Woods et al., 2001; FDA standards). Moreover, approaches to quality improvement in healthcare emphasize human factors as one of the key perspectives to take in driving quality (e.g., Hughes, 2008; Ting et al., 2009; Varkey et al., 2007).

By considering human capabilities in relation to the demands of the task, human factors can play a key role in identifying problems and offering concrete solutions. By that, we can mitigate risks and contribute to continuous quality improvement. This is what this book is about.

1.3   ABOUT THE BOOK

We designed this book, *Human Factors in Healthcare: A Practical Guide to Quality Improvement*, with our own human limitations in mind. Knowing we aim to improve work processes and technological solutions that are increasingly complex, knowing that it is impossible to take all the factors into account all the time, and knowing the serious consequences that missing critical factors can have on the outcomes (such as patient safety), we designed a process that helps mitigate our limitations using a systematic approach. This approach, aligned with other quality improvement frameworks (such as that of the Agency for Healthcare Research and Quality, 2013, which includes the Plan-

Do-Study-Act Model for Improvement, Langley et al., 1996), can help you consider the fuller picture and ultimately achieve better results.

1.3.1   TARGET AUDIENCE

If you are involved in any of these roles, you can benefit from this book.

- Quality manager: monitoring and advising on the performance of the quality management system, reporting on performance, and measuring against set standards.

- Risk manager: assessing, identifying, and mitigating risk.

- Process engineer: assessing and improving processes.

- Regulator, adverse events auditor, assessor, investigator: identifying potential adverse events and reporting on the risk.

- Architect or industrial designer: designing industrial products, interiors, and even architecture.

- Human factors engineer: designing new and improving technology products, decision-making processes, work settings, and living environments.

While the book focuses on healthcare, the methodology is appropriate for analyzing human factors in other complex environments as well.

1.3.2   A DIFFERENT KIND OF A HUMAN FACTORS BOOK

This book is different from other books on human factors, human performance, and human-machine interface design. In this book, we guide you through the human factors analysis process in order to take action.

Consider It a Workbook and a Field Guide:

Use it—don't just read it! It is a guide through a process. Use it as a guide to approach, analyze, and solve complex problems in healthcare systems where people and technology are involved. Apply the methodology to each case. As you analyze each case, you will find increasing similarities but the differences may be important as well.

Use It for Proactive and Reactive Continuous Improvement

The process provides tools to frame the problem or situation in ways that make it possible to use the information to reactively solve known problems. It addresses the complex interrelations among the relevant factors.

It can also be used proactively before problems arise. This can be for the purpose of improving the current situation to make it even better or designing new solutions. It can be in the planning stages of something new, such as a new system, environment, or change.

What this Book Is Not…

While you can learn about human factors engineering from this book, this is not a scientific or academic textbook. You will not find extensive and comprehensive reviews of the relevant literature here. If you want to learn the fundamentals of human factors engineering, in general and specifically in healthcare, use this book as a methodological supplement to other textbooks (that we will reference later).

Moreover, while this book guides you to analyze the situation and propose solutions for continuous improvement, it is not a human factors engineering design guide. We refer you to other sources and resources that you can use as a design guide to complement your proposals for interventions and mitigations.

1.3.3   BOOK ORGANIZATION

The book is divided into two major parts.

1. A Conceptual Framework. The first part is a relatively short conceptual overview of the human factors framework and the accompanying practical methodology to analyze situations, events, sites, and processes.

2. The Field Guide. The second part is a step-by-step outline of the methodology you can use to analyze and solve problems using a human factors approach.

We recommend that the first time you read the book, begin with the Conceptual Framework before going on to The Field Guide. Then each time you need to analyze a situation, use The Field Guide as a performance support tool to systematically cover as many factors as appropriate.

PART I

A Conceptual Framework

In this part of the book, we explain the rationale for the Human Factors Conceptual Framework to Map-Assess-Recognize-Conclude (HF-MARC).

- In Chapter 2, "About Human Factors Frameworks," we discuss:

  ○ the rationale for using a framework and

  ○ explore a number of existing frameworks for human factors analysis in healthcare.

- In Chapter 3, HF - MARC: we introduce the HF-MARC framework that builds on existing frameworks, synthesizing them with best practices to offer an integrated methodology for analyzing, mitigating, and improving healthcare from a human factors perspective.

The second part of the book actually walks you through the process of using the HF-MARC.

CHAPTER 2

About Human Factors Frameworks

The complexity and criticality of healthcare environments involve many factors, including the humans involved, the environment in which they work, and the tasks they perform. By analyzing the human factors involved, it is possible to improve the tools, processes, and environment layout, ultimately improving healthcare outcomes, such as patient safety. Intuitively, we can identify many human factors that might involve risks and opportunities for improvement. At the same time, with a systematic approach, we believe it is possible to achieve even better results. In the context of healthcare, these improved results could ultimately save lives.

2.1 WHY HAVE A FRAMEWORK?

By definition, a framework supports a systematic approach. A framework:

- simplifies, explains and shows inter-relations;

- serves as an orientation map; and

- provides the freedom to ask questions and search for the answers within its structure.

In healthcare, a human factors framework provides:

- tools for designing and redesigning healthcare systems and processes in order to improve patient safety and care quality; and

- a structured approach for handling information that turns overload of chaotic information into meaningful information you can use.

2.2 FRAMEWORKS OF HUMAN FACTORS IN HEALTHCARE—A REVIEW

There are a number of human factors frameworks focused on healthcare or relevant to it. Let us review some of them very briefly. Each framework looks at healthcare somewhat differently.

2.2.1 THE SEIPS MODEL—SYSTEMS ENGINEERING INITIATIVE FOR PATIENT SAFETY

The SEIPS model (Carayon et al., 2014b) looks at the work system model, patient, employee, and organizational outcomes. With the person at the center of the work system, it looks for feedback loops from processes and outcomes and how the process is influenced by the work system. This model has been used to study and improve the practice of healthcare. It looks at what processes are involved and how they lead to patient, employee, and organizational outcomes that act as feedback to the different players within the external environment (see Figure 2.1).

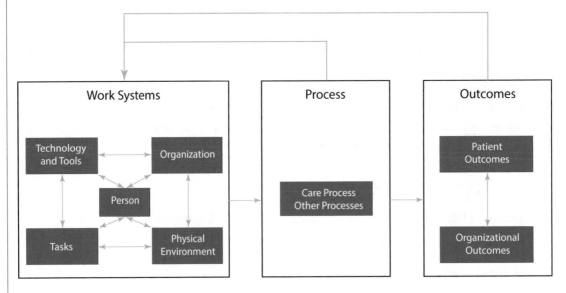

Figure 2.1: A generalized representation of the SEIPS model (adapted from Carayon et al., 2014b).

Key characteristics (as described by Carayon et al., 2014) include the following:

1. **Description of the work system and its interacting elements**, including internal staff (doctors, nurses, other staff, patient-centered medical home team, patients, families), and external environment (regulatory, professional and consumer/patient groups in healthcare delivery).

2. **Incorporation of quality of care model:** the "Structure–Process–Outcome" approach of Donabedian (1978) (Donabedian: material resources (e.g., facilities, equipment), human resources (e.g., number and qualifications of staff), and organizational structure (e.g., organization of medical staff, methods of reimbursement), plus a systematic approach to looking at the work system (replacing Donabedian's structure).

3. **Identification of care processes influenced by the work system and contributing to outcomes:** in conjunction with total quality management and lean thinking, this set of characteristics helps ensure that process analysis and quality improvement efforts in healthcare integrate human factors engineering (HFE) issues.

4. **Integration of inter-related patient outcomes and organizational/employee outcomes** (health or disease outcomes, personal satisfaction, quality of life, financial solvency). The goal of healthcare system (re)design is to benefit patients and healthcare workers and organizations. Therefore, the changes should benefit both patients and healthcare workers.

5. **Feedback loops between process and outcomes, and the work system:** by collecting data on care processes and outcomes, including the patients, employees, and organization, problems can be identified and used in improving the design of the work system.

2.2.2   HUMAN FACTORS ENGINEERING PARADIGM

The HFE paradigm to patient safety, proposed by Karsh et al. (2006), takes a systems approach to assessing human factors. It looks at any given situation or event in terms of performance inputs, transformation processes, and performance outputs. Each of these factors covers a variety of factors all of which are reviewed in assessing healthcare professional performance. Within the input factors they list the patient and healthcare professionals, the task, technologies and tools, environmental factors, organizational factors, and other factors external to the immediate environment. Within the transformational processes they list factors such as cognitive, action, and social activities performed by people. Within the output factors they list changes that may occur in the environment, the task, or the people themselves, and the clinical outcomes (such as surgery completed). The outcomes part of the outputs in this system is evaluated in terms of meeting objectives and standards. If those are not met, the system must be redesigned.

2.2.3   A HUMAN FACTORS FRAMEWORK FOR ANALYZING RISK AND SAFETY IN CLINICAL MEDICINE

The framework suggested by Vincent et al. (1998) emphasizes the human factors in adverse events. The overall approach is based on Reason's Swiss cheese metaphor with latent factors, work conditions that moderate the impact of the latent factors, and active failures. Among the various factors, at the different levels of the process, they list work and organizational factors, individual and team factors, task characteristics, and finally, of course, patient factors.

2.2.4   THE FAA'S HUMAN FACTORS ANALYSIS AND CLASSIFICATION SYSTEM (HFACS)

The Federal Aviation Administration's Human Factors Analysis and Classification System (HFACS) is also aimed primarily at the analysis of accidents. While the framework guides toward factors that are unsafe, the aspects relevant to us are what they refer to as the preconditions for unsafe acts. Those conditions are factors associated with the condition of operators and factors associated with practices and actions of the operators. From those we can derive factors such mental and physical state of the operators, and how they manage their actions.

2.2.5   THE WHO HUMAN FACTORS FRAMEWORK

WHO (2009) first looks at the patient in the work environment, including the equipment; it then looks at the individual, the team/group, the organization and management and, finally, societal, cultural, and regulatory influences. For each of these factors, it includes a number of factors to be considered:

- work environment—hazards;

- individual;

    ◦ cognitive skills (situational awareness, decision making);

    ◦ personal resources (stress, fatigue);

- workgroup/team;

    ◦ teamwork (structure, processes, dynamics);

    ◦ team leadership (supervisors);

- organizational/managerial functions;

    ◦ safety culture;

    ◦ manager's leadership;

    ◦ communication.

2.2.6   THE FOOD AND DRUG ADMINISTRATION HUMAN FACTORS FRAMEWORK

The Food and Drug Administration's (FDA)'s human factors framework focuses on device use safety in the context of risk management. The framework suggests three categories of factors: (1) the environment itself; (2) the user; and (3) the device. The environmental factors include physical aspects such as noise and lighting. The user factors include knowledge, abilities, and limitations. Finally, the device factors include aspects such as the user interface, operational requirements, and complexity.

2.2.7   A SUMMARY COMPARATIVE TABLE

Table 2.1 shows commonalities and differences among some human factors frameworks. You can use most of them to analyze both proactive cases—finding opportunities to improve; and retrospective cases, including accident and adverse event analysis. There are also frameworks aimed only for retrospective cases.

2.3   CRITIQUE AND SUMMARY

Analyzing a case with the guidance of a framework is useful. Given that there are several frameworks relevant to healthcare, how would you know which one to use? Here are some questions that guided us.

- Is the framework adequate theoretically? In other words, can the framework suggest an explanation for what happened or for what may happen?

- Importantly, is the framework useful as a practical guide? In other words, does the framework offer a clear, step-by-step methodology for a human factors analysis?

Learning from other frameworks as well as from our experience, study, and research in the area (i.e., Parush et al., 2011a), this book offers a framework that can elicit explanations for what happened or may happen. Its rationale is based on the complex interaction between people's abilities and the demands of the tasks they do; and the extent to which the environment supports success. Moreover, the framework provides a detailed step-by-step methodology for assessing and improving complex, mission-, and safety-critical work environments in healthcare.

Table 2.1: Comparison and summary of human factors frameworks

Factors	Framework Aim					
	Proactive and Retrospective Modeling and Analysis				Accident and Adverse Event Analysis	
	WHO 2009	DFA	HFE Paradigm (Karsh et al.)	SEIPS: HF systems approach Carayon et al.	FAAs HFACS	Framework for analyzing risk and safety in clinical medicine
Environment	Work environment; Societal cultural and regulatory influences	Use environment; light; noise distraction; motion/vibration	External environment	External environment		
Device	Equipment	Operating requirements; procedures; complexity; UI characteristics	Work system/unit factors	Part of technology and tools		Work environment
Individual	Individual	Users: knowledge; abilities; expectations; limitations	Patient and provider factors	Patient outcomes; employee outcomes	Substandard conditions of operators; unsafe supervision	Individual staff factors Patient characteristics
Organization Factors	Organizational and management factors		Organizational factors	Organizational outcomes		Institutional context; organizational and management factors
Processes			Transformations	Processes	Substandard practices of operators	Task factors
Group/Team	Team (group)					Team factors

CHAPTER 3

HF–MARC:

Using the Human Factors Conceptual Framework to Map-Assess-Recognize-Conclude

3.1 OVERVIEW

Healthcare organizations and cases are complex.

What do we mean by complexity? Something that is complex has some of the following characteristics.

- Many factors are influential. Traditional root cause analysis is often used to identify multiple factors leading to an adverse event.

Other aspects of complexity fit less neatly into a traditional root cause analysis, including the following.

- The influence of the various factors is interwoven and inter-dependent.

- The influence is holistic, systemic. It is not only about people. It is not only about the environment. It is about the combined influence of all of them together.

- The spread of influence is nonlinear. That means that the size of each influencing factor does not determine its impact—something small might have a big impact; something with a large input might have a small influence.

- There are fuzzy boundaries between the factors and their influence.

Using a guiding framework makes it possible to understand what happened or what may happen in them, making the complexity more manageable. In order to mitigate the factors contributing to health and safety risks in healthcare systems, we need a framework that can provide a methodology to uncover the complexity and to guide us through an effective analysis and problem-solving process.

This book offers a conceptual framework (originally introduced briefly by Parush et al. (2011a) that supports discovering and improving the human factors that may influence each patient-care chain-of-events. The framework offers a methodology with a solid conceptual foundation, and thus has two components:

1. **HF** Conceptual Pyramid: a structure of influencing factors that result in outcomes, including the extent to which the goals are met

2. **MARC** Methodology: the Map-Assess-Recognize-Conclude (MARC) process

The objective of the human factors conceptual framework and the MARC analysis methodology presented in this book is to guide you in analyzing and discovering what influences the performance of people, organizations, and systems and brings about certain outcomes, and then draw conclusions regarding relevant interventions and mitigations.

3.2   HF CONCEPTUAL PYRAMID: RATIONALE

The human factors framework tells a "story" about a person, team, or organization; their goals and the tasks they perform to achieve them. The success or failure in performing tasks depends primarily on the simplicity or complexity of the tasks, and the qualifications and experience of the people performing them. People perform tasks in a certain context, which can also influence their success or failure. One aspect of the context is the physical environment. Another aspect of the context includes other people and the organization within which they perform. As long as the context is supportive, the task is not too complex, and the people are qualified and experienced; it is highly likely they will succeed. However, if the task is too difficult or if the context is disruptive—then the people will have a hard time coping with those obstacles. Because of the tension between the nature of the *task* and the *environment*, on the one hand, and what *people* can cope with, on the other, things may escalate further and *new aspects can emerge*. For example, people may also get stressed. All of these together influence performance of the task and the consequences of that performance.

Now, let us formalize this story. The fundamental conceptual premise of the HF-MARC framework is that a variety of tightly inter-related factors can influence the manner in which people, organizations, and systems perform tasks and achieve their goals. These factors are:

- people;

- tasks;

- environment—the physical and the human aspects of the environmental context within which people perform their tasks;

- capabilities and limitations of the people; and

- emergent factors—new factors that arise as a result of the interaction between other existing factors.

All of these factors have a combined impact on the performance (the actual behaviors of the key players) and outcomes (the extent of success or failure). The framework is presented graphically in Figure 3.1.

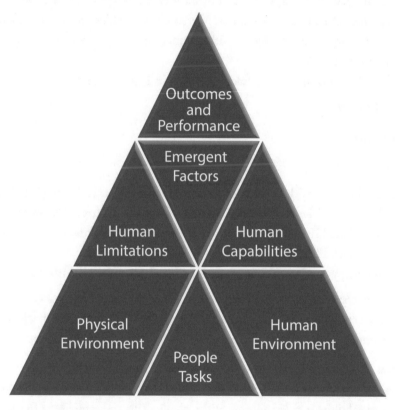

Figure 3.1: The conceptual part of the HF framework that expresses the systemic relations between various factors.

We use a pyramid visualization with three tiers to schematically represent the factors and their inter-relations. Let us walk through this visualization from the bottom up. The foundation tier represents the context: the people, their goals and tasks, and the environment—physical and human—in which one performs the tasks. The top tier represents the performance of the task and the outcomes. In between the foundation and the top tiers, there is a middle, moderating tier: the one consisting of the human factors—human capabilities and limitations and the factors emerging from the tension between the task and environmental demands on the one hand, and human capabilities and limitations on the other.

3.3   METHODOLOGICAL PROCESS: RATIONALE

You can use the framework to explain phenomena between people and their work environment, and the resulting performance and outcomes. Moreover, it includes an inherent methodology for analyzing situations and solving problems with a human factors perspective. As such, that framework is really a practical tool.

The basic rationale of the methodology is a simple 1-2-3 process.

1. Collect the facts.

2. Assess them.

3. Draw conclusions.

We created the framework by taking this simple approach and adapting it to incorporate the HF Conceptual Pyramid. The methodological part of the HF-MARC framework translates the conceptual aspects of the HF pyramid into a series of practical questions, starting from the bottom and moving upward (see Figure 3.2).

The methodology consists of four research and analysis activities with the acronym MARC.

- **Map:** This corresponds to the first step: "Collect the facts." Mapping is the detailed outline of various aspects of the tasks and the environmental context in which the tasks are performed. This corresponds to the foundation tier of the HF conceptual pyramid.

- **Assess:** This corresponds to the "Assess the facts" step above. Assessing is the detailed analysis of the fit between (1) the demands of the task and environment and (2) the human capabilities and limitations of the people. This corresponds to the moderating tier of the HF conceptual pyramid.

- **Recognize:** This step builds on assessing the facts. Recognizing is the additional analysis we perform to discover factors that emerge as a result of the interaction (fit and lack of fit) between task and environmental requirements on the one hand, and human capabilities and limitations on the other. This is part of the moderating tier of the HF conceptual pyramid.

- **Conclude:** Drawing conclusions regarding human performance and outcomes, and implications for mitigation and risk management. This corresponds to the top tier of the HF conceptual pyramid.

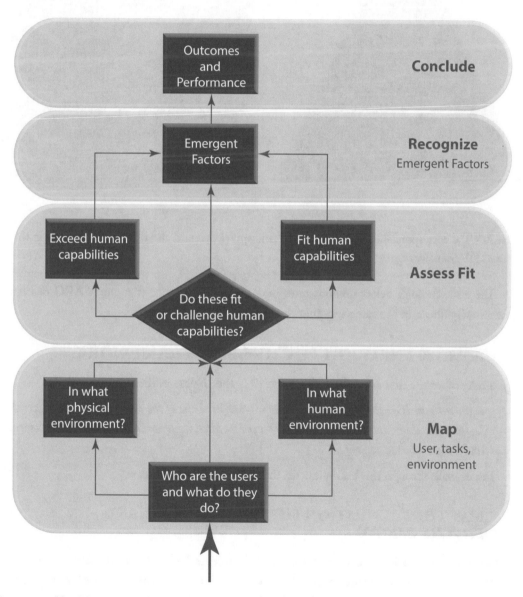

Figure 3.2: The Map-Assess-Recognize-Conclude (MARC) process.

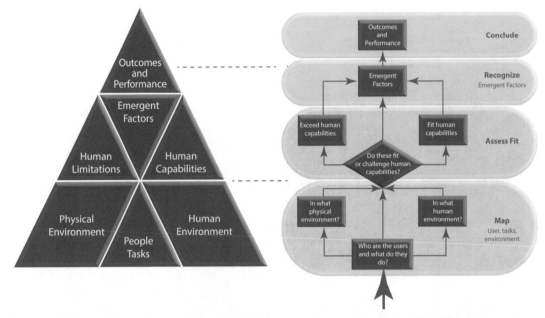

Figure 3.3: The correspondence between the HF conceptual pyramid and the MARC process in the proposed HF framework.

The walk-through below demonstrates how this works. Notice that the MARC process is not necessarily linear, in the same way that the framework is not linear.

3.4   A WALK THROUGH THE HF-MARC FRAMEWORK

Do you remember the first example in this book? It is the "Nurse in the ICU example."

A nurse in an ICU walks over to check the vital signs of one of the patients. The nurse succeeds in this task: she notices all the parameters on the monitor, understands the status of the patient, and knows what to do next.

Let us walk through this example with the framework as a guide.

3.4.1   MAP THE FOUNDATION TIER: PEOPLE, TASKS, AND ENVIRONMENTS

The foundation tier consists of people, tasks, and environment. Here we map the facts that describe them. We start with people. People are at the core of any human factors work because we are primarily concerned with fitting what we design and provide to the people who will use it.

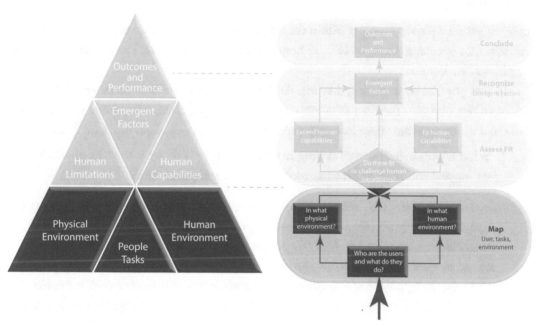

Figure 3.4: Mapping the factors in the foundation tier: people, tasks, and environments.

People and Tasks

The first mapping question we should always ask ourselves is:

1. **Who are the people involved in the site, situation, or process** we want to analyze?

   In the "Nurse in the ICU" example, the simple answer to this question would be the patient and the nurse. An additional aspect is that the nurse has experience working in the ICU. We will continue focusing on the nurse because our analysis is concerned with factors that influence the nurse's success.

People have goals. People do things to achieve those goals. People have experience as part of doing (or not doing…). The second set of questions we should ask is:

2. **What tasks do the people do?** What do they want to do? What are they required to do? **When** do they do them? **How** do they do them?

   In the "Nurse in the ICU" example, the simple answers to those questions would be to check the vital signs monitor (the "What"), every 10 min (the "When"), by walking over to the patient's bedside monitor and looking at it (the "How").

Environment

People act within a physical context within the site, situation, or process we are analyzing. It could be an office, an operating room, an ICU, at home, in a vehicle, and more. This context constitutes their physical environment.

In that physical environment, they can do things as individuals or as members of a group, a team, or an organization. They can follow certain routines, regulations, rules, protocols, and more. All of those would constitute their human environment. The third set of questions we should ask is:

3. **What is the environmental setting** in which the people do what they do, physically and organizationally? **With what tools—physical and procedural?**

   *In the "Nurse in the ICU" example, the simple answers to those questions would be in the ICU (a physical and organizational **Environment**), part of the usual ICU team (the organizational **Environment**), in a hospital (the physical and organizational **Environment**), using the bedside vital signs monitor (the physical **Tool**). We can add that the ICU is mostly quiet with adequate lighting, and that there is a clear and direct path from the nurse's seat to the beds of the patients she monitors and handles. Figure 3.5 shows a very simplified view of how some factors in the Nurse in the ICU example are mapped with the foundation tier.*

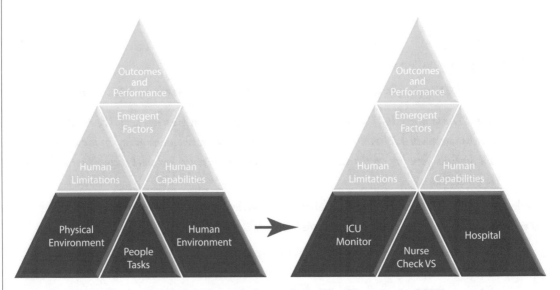

Figure 3.5: Mapping the foundation tier with factors from "The Nurse in the ICU" example.

Taken together, the answers to these questions help map the framework's foundation tier—the people, their goals and tasks, and their environments.

3.4.2   THE MODERATING TIER: ASSESS THE FIT AND RECOGNIZE EMERGENT FACTORS

In the moderating tier, as part of "assessing the facts", we first map the relevant human capabilities and limitations. Then we recognize factors that emerge as a result of the interaction (fit and lack of fit) between task and environmental requirements on the one hand, and human capabilities and limitations on the other.

Assess Fit to Human Capabilities and Limitations

As human beings, we can do a lot. We have well-developed intellectual capacities, as well as physical capabilities. We can learn, pay attention, remember, think, and make decisions. We can move and lift things. Our abilities are influenced by internal factors as well as by the context. For example, our ability to complete a task successfully may be influenced by the surrounding noise, number of interruptions, and other factors that arise from the context. When we succeed, we feel satisfied, happy, and delighted. We can even deal with stressful challenges beyond our expectations. Nevertheless, sometimes we fail because we are also limited in what we can do. We can miss things in our vicinity and we can forget. We can feel sad, frustrated, and annoyed. By examining people's capabilities and limitations; as well as the extent to which the environment supports or hinders those capabilities and limitations, we can gain insight about how and why people, teams, organizations, and systems succeed or fail in their tasks.

The fourth question we should ask is:

4. **What are the relevant human capabilities and limitations? Do they fit the tasks and demands of the environment?**

   Relevance is critical. Here we are looking for the human capabilities and limitations that relate to the given context with respect to the people, their tasks, and the environments. When we talk about capabilities and limitations, we consider cognitive, emotional, and physical aspects.

   In the "Nurse in the ICU" example, the answer to those questions would be the nurse's capability to visually perceive and understand the information displayed on the vital signs monitor. We could deduce this understanding based on having determined that the nurse is trained, qualified, and experienced in reading vital signs and understanding their meaning, when answering the Who question.

However, if there was an additional twist in the case whereby the nurse was busy with several patients while overseeing a new nurse in preparing medications, then we would point out some limitations such as the limited ability to divide attention and multi-task.

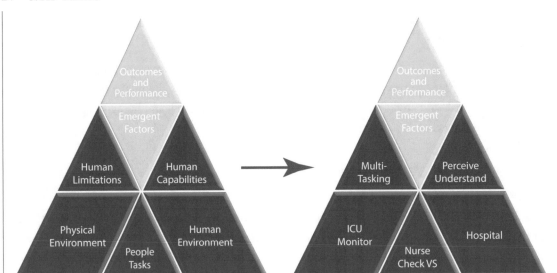

Figure 3.6: Assessing the fit with human capabilities and limitations in the moderating tier of the framework.

Recognize Emergent Factors

Taken together, the human capabilities and limitations make up a major part of the middle tier of the HF Conceptual Pyramid. However, that is not all. There are other kinds of human factors that emerge from the tension between the demands of the tasks and the environments on the one hand, and human capabilities and limitations, on the other hand. For example, the task may be very demanding both cognitively and physically, and as a result we may feel stressed. That stress is a new emergent factor.

The fifth set of questions we should ask is:

5. Can we recognize any **factors that emerge** because of the tension between task and environment demands, on the one hand, and human capabilities, on the other hand? Are capabilities exceeded?

   In the "nurse in the ICU" example, the answer to this question would be that there are no negative emergent factors, such as stress. This is because the demands of the task—to look at the vital signs monitor and understand what the vital signs mean—did not exceed the nurse's capabilities. It also seems that the environmental conditions did not present any challenging demands.

However, an additional twist in the story in which the nurse was busy with several other patients while overseeing another nurse who was preparing medications—would create a greater tension between the task and environmental demand, on the one hand, and capabilities on the other. Such tension may result in stress, which we would consider an emergent factor in this framework.

The middle tier of the pyramid is considered "moderating" because these human factors, including the basic **human capabilities and limitations** as well as the **emergent factors**, moderate **people's** final performance of the **tasks** given the **environmental** context, and thus influence the **outcomes**.

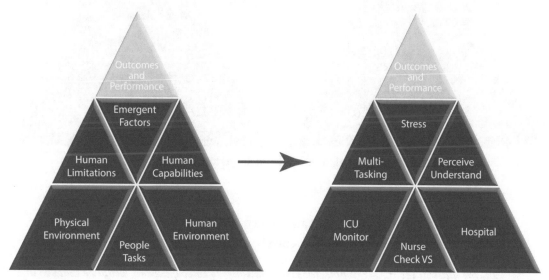

Figure 3.7: **Recognizing the emergent factors that result from the tension between demands and capabilities.**

3.4.3   THE TOP TIER: CONCLUDE ABOUT PERFORMANCE AND OUTCOMES

To recap, the fundamental premise of the HF-MARC Framework is that a variety of tightly inter-related factors can influence the manner in which people, organizations, and systems *perform* tasks and achieve their goals.

We use the HF Conceptual Pyramid and the MARC analysis methodology in order to draw conclusions regarding what influences the performance of these people, organizations, and systems, and what brings about certain outcomes.

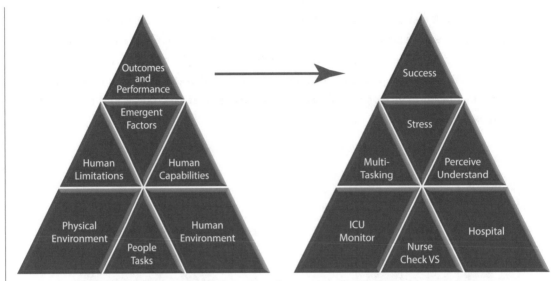

Figure 3.8: Concluding about individual, team, and system performance and outcomes.

The sixth question we ask is:

6. How does the inter-play between all the **factors influence performance**? How does this performance **influence the outcomes**?

In "the nurse in the ICU" example, the simple answers to those questions would be that the inter-play between the factors had a positive influence, and the tasks were completed successfully. That means the clinical outcomes were positive as well: the patient was attended on time, the nurse knew what was going on, and subsequent treatment was adequate. But consider the bad twist in the story again. It is possible that vital signs were not checked on time, or the nurse did not get all the important information, and perhaps the nurse was not aware that the patient's status is deteriorating. That could have led to harmful outcomes such as delayed treatment or severe status deterioration.

3.4.4   INTERVENTIONS AND MITIGATIONS

The ultimate goal of understanding what factors influence performance and outcomes is to decide what to do about them. We aim to preserve and strengthen the positive influences, on the one hand, and mitigate the negative influences, on the other.

Notice that interventions and mitigations build on the framework's conclusions. The conclusions of the process define needs. The final set of questions focus on the ways we might address those needs. Here we ask:

7. What can we do in order to **facilitate and preserve good performance** and outcomes? What can we do to mitigate negative influence on performance and **prevent harmful outcomes**?

In the "nurse in the ICU" example, the simple answers to those questions would be that we want to keep the environment quiet so no distractions or interruptions will occur, to ensure that the nurse's workload will not exceed the maximum number of patients she can handle, and not clutter the environment with more beds or devices so as to keep the path clear. If we look at the possible negative side of the story, then we could prioritize mitigating the influencing factor of increased workload by not having to oversee the work of other nurses while she is in charge of attending to several patients, and allocating a separate and quiet physical space for her other work.

PART   II

The Human Factors Field Guide

CHAPTER 4

Overview: A Human Factors Approach to Continuous Improvement

This part of the book is designed as a job aid, to support the process you use to analyze adverse events and propose improvements in the system. It is designed as a guide to continuous improvement.

4.1 RECAP OF THE FRAMEWORK, METHODOLOGY, AND A ROADMAP

Here is a very brief reminder of the framework in Figure 4.1. The MARC process on the right is the process we are about to follow step-by-step. Table 4.1 summarizes the key questions we ask with every step in the MARC process. Table 4.2 shows a roadmap of the entire process.

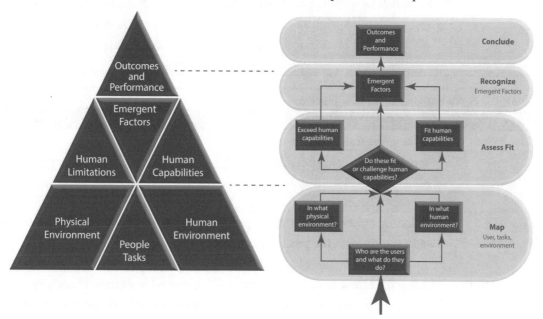

Figure 4.1: The HF Conceptual Pyramid and the MARC process are aligned in the proposed human factors framework.

Table 4.1: Summary of the key steps and questions in the MARC methodology

MARC	The Key Questions
Map	1. Who are the people involved in the site, situation, or process we want to analyze?
	2. What tasks do the people do? What do they want to do? What are they required to do? When do they do them? How do they do them?
	3. What is the environmental setting in which the people do what they do, physically and organizationally? With what tools, physical and procedural?
Assess Fit	4. What are the relevant human capabilities and limitations? Do they fit the tasks and demands of the environment?
Recognize	5. What factors that emerge can we recognize because of the tension between task and environment on demands, on the one hand, and human capabilities, on the other? Are capabilities exceeded?
Conclude	6. How does the inter-play between all the factors influence performance? How does this performance influence the outcomes?
	7. What can we do in order to facilitate and preserve good performance and outcomes? What can we do to mitigate negative influence on performance and prevent harmful outcomes?

Table 4.2: A roadmap of the MARC process

Chapter	Walks you through the process of:
Overview	Viewing the framework at-a glance.
Start the Process	Planning your project by identifying the goals and objectives as well as the scope of the project. Based on these, as well as other constraints, you can choose a lean or detailed strategy.
Map the context	Collecting information about the people, tasks, and environment. This guide walks you through the process using the detailed checklist.
Brief # 1	Summarizing the key issues and presenting to stakeholders for validation
Assess the fit	Rating the extent to which the players within the relevant units of analysis (individual, group, team, department, and organization) fit the context and the context fits the people. At this stage, problems and influential factors emerge.
Interim findings: problems of fit	Are there interim findings that require immediate attention?

Chapter	Walks you through the process of:
Recognize emergent factors	Looking for additional factors that emerge as a result of the fit between the demands of the task and environmental factors, on the one hand, and human capabilities, on the other. These may have a key impact on performance and outcomes.
Brief # 2	Summarizing the key problems and presenting to stakeholders for validation
Conclude :	About performance and outcomes, explaining what happened or what may happen and what should we do about it—defining the needs.
Interventions and mitigations	Determining the relevant interventions and mitigations to improve performance and outcomes.
Brief # 3	Summarizing the most influential factors, performance and outcomes, and proposing interventions and mitigations to stakeholders.

4.2   THE SAMPLE SCENARIO

In this part we will use the following scenario as an example in walking you through the process. All instances of the sepsis scenario will appear with the bluish background.

Sepsis Management Scenario

Heather, a 47-year-old woman, is quite healthy at baseline. She has had several episodes of renal colic (flank pain from kidney stones) throughout her life, and underwent laparoscopic removal of a large right-sided kidney stone 4 or 5 years ago.

Heather came to the Emergency Department at 6:30 AM after experiencing the too-familiar right-sided flank pain for several days. She entered the Emergency Department's main entrance and somehow managed to pull herself over to the registration cubicle. She was somewhat confused, and was not able to remember her home address or phone number. Her first recorded vital signs were as follows: HR, 128; BP, 100/65; Temp, 38.8 0C; RR, 24; O_2Sat, 94% on room air.

Heather ended up waiting for 45 min before she was registered. The triage nurse did not screen her for sepsis, and did not appreciate that she was very ill. She ended up staying in the waiting area for an additional 1.5 hr, receiving no care at all. She then collapsed—fell on the floor, lost consciousness—in the waiting area, and this prompted the staff to bring her urgently to one of the resuscitation rooms in "Section A." At this stage she required aggressive resuscitation, including intubation (of the windpipe), mechanical ventilation, urgent placement of central venous catheter, and admission to the ICU. It took 4 hr before the first dose of antibiotics was given, and blood cultures were not drawn at all. The delay in managing her resulted in acute kidney failure, and she required hemodialysis for several weeks. She ended up requiring critical care for 3 weeks, then 2 additional weeks on a regular hospital ward, and finally rehab at the local center.

CHAPTER 5

Start the Process

You should start the process by considering and making basic decisions that will guide later considerations. Those basic decisions address the goals and objectives of your analysis and the key stakeholders in the context. Following that we introduce the checklists that you can use for the analysis.

5.1 DETERMINE OBJECTIVES AND SCOPE OF THE ANALYSIS

This section talks about three very important aspects you must determine at the beginning:

- objectives of the analysis;

- the unit of analysis; and

- the relevant stakeholders.

Ask yourself: Why am I doing this analysis? What is the **objective of the analysis**? Here are some possible answers:

- process or quality improvement;

- part of redesigning a process or a site;

- part of a new technology deployment; and

- analysis of a near-miss or adverse event.

Also, ask yourself what is the **unit of analysis**? Here are some possible answers.

- An **individual:** each stakeholder in the scope of analysis can be analyzed as an individual.

- A **team:** the stakeholders can also be seen within the context of a number of people collaborating or connected in terms of achieving their purpose. Examples could be the hospital staff performing an operation or on duty in an emergency during a single shift. It could also be a group of staff working together on a project, like reviewing a new technology to recommend a procurement decision.

- An organizational **unit:** the stakeholders often are part of a department, division, or branch within an organization. As such, they may share a variety of conditions and experiences, including a common purpose, set of skills, group dynamics, and unique challenges.

- A whole **organization** or site: while each unit, team, and individual has its own unique characteristics; there are some factors that are shared by everyone present within the whole organization or site. It may be important to look at the organization as a whole, both on a human level—with its culture, values, traditions, policies, and processes—and on a physical level—with its layout, light, noise, space, furniture, and tools.

- A **system**: using a systemic lens for analysis allows you to see the interactions among participants from the perspective of inputs, processes, and outputs. It encourages tracking the information flow, surfacing the feedback mechanisms that modulate the behavior and consequent performance and outcomes.

Finally, identify the *stakeholders*, the target audience for the analysis. Those are the people or organizational units that will receive the report of the analysis and implement its conclusions and recommendations. The identification of the stakeholders is tightly linked with the objectives of the analysis. By producing and presenting briefs on the progress of your analysis, you involve the stakeholders throughout the process, give them an opportunity to influence and validate, and get an early buy-in that will facilitate providing recommendations that will be likely implemented.

The following is an example of objectives and the unit of analysis for the sepsis management scenario.

Objectives—Analysis of an Adverse Event

We want to understand why the patient was not screened for sepsis, what were the causes for delays throughout the management of the case, and why did the case end up with outcomes that included resuscitation, ICU, hemodialysis, and prolonged critical care?

The Units of Analyses

The ED team at the time with a focus on the triage nurse

Stakeholders

ICU directors, hospital management

5.2   USE CHECKLISTS FOR THE MAPPING AND ANALYSES

As you saw, the MARC process is about questions. The analysis work is really about asking all the right questions. Or, more accurately, you methodically ask all the questions, determine which of the questions are relevant to your situation, and then seek the answers to those questions. One great way to methodically ask all the relevant questions is to use a checklist. This book offers you a choice between a short and a detailed checklist, depending on the type of analysis you require and the time you have available. Where appropriate, we offer additional analysis tools that offer insights by synthesizing information from the answers to the above questions, such as link analysis.

5.2.1   THE SHORT MAPPING CHECKLIST

As the name implies, this is a short list of questions that aids you during the mapping phase of the MARC process. That is, it helps in mapping the foundation tier in the conceptual framework, the key factors in the context. The short checklist could be very useful in data collection (for example, during observations and interviews, or collecting data from secondary sources). Because it is a very short list of questions, it will guide you to collect data that will be highly relevant later in the detailed mapping and assessment steps of the analysis.

The short mapping checklist is composed of five categories.

1. **People:** who they are, what their goals are, and what their relationships are.

2. **Tasks:** what they do, what they are supposed to do, and how.

3. **Physical environment:** where they do their tasks and with which tools.

4. **Human environment:** in which organization, alone or in a team, with what policies.

5. **Behaviors, performance, feelings, impressions:** Additional observations

Figure 5.1 shows a short checklist.

HUMAN FACTORS MAPPING CHECKLIST

People
    Who are the people?
    What are their goals and roles
    What are the relationships among them?

Tasks
    What do they do?
    What are they supposed to do?
    How do they do it, or supposed to?

Physical Environment
    Where are they doing their tasks?
    How is that place organized?
    What tools do they use? Where are they located in the place?

Human Environment
    In what organization do they do the work?
    What is its structure?
    Is there a group or teamwork?

Behaviors, Performance, Feelings, Impressions
    What other observations do you have?

Figure 5.1: The short human factors mapping checklist.

5.2.2   THE DETAILED CHECKLIST

The detailed checklist is aimed at guiding and supporting all phases of the MARC methodology. So even if you did the mapping phase with the short checklist, you can then proceed with the detailed one to guide you in analyzing the fit between the contextual factors, and human capabilities and limitations. It broadens the analysis by asking questions that highlight emergent factors, and lead to meaningful conclusions.

We will use the detailed checklist in all the following sections since it provides the most detailed approach to the analysis and in following the conceptual framework. You can find the entire detailed checklist in Appendix A.

5.3   CHOOSE THE CHECKLIST FOR THE ANALYSIS

Given the objectives you determined for your analysis based on a review of the scenario and units (such as people, teams, organization, and system) involved, which checklist do you feel is appropriate to use?

- The short list lets you dive into the first tier questions.

- The detailed checklist takes you systematically through the process, with a wider and deeper scope of questions.

As you gain experience, you may choose to begin with the lean checklist and then expand in the areas of interest using the detailed one. Whichever approach you choose, determine which of the questions are relevant to your situation by referring back to the objectives you set and units you chose in determining the scope of the analysis. While these definitions may change as your analysis progresses, it is important to update and maintain focus on your goals in order to achieve them.

CHAPTER 6

Map the Context

Mapping the context has to do with collecting data and information about the context where people perform their tasks. In some cases we perform additional analyses to understand the raw data more deeply and represent it more schematically (such as task analysis), and sometimes we just leave the raw data and use it, as is, later in the fit assessment phase.

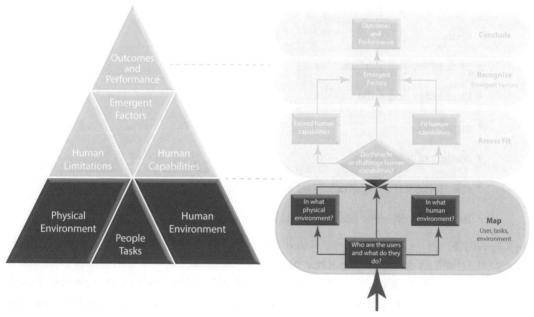

Figure 6.1: Mapping the context.

6.1 PEOPLE

The first step is to map the people who are the primary and additional stakeholders in a given context, defined by place and time. The primary stakeholders can be patients and their families, healthcare providers, or anyone involved directly in the healthcare process. It may be interesting to see the people in the context of a social network, mapping the interactions and considering the level of involvement of each person. Later in the process, we will expand the scope of mapping people toward the broader human environment (which focuses more on organizational aspects).

Stakeholders can be "profiled" according to factors such as roles (for example, patient, family, and care provider), professions (for example, nurses, surgeons), knowledge and experience in the relevant profession, context (for example, surgery, ICU, and community clinic), and by level of involvement or influence (for example, a surgeon in a surgery is more directly involved than the hospital administrator). In the context of healthcare, it is important to remember the patients!

Use the detailed checklist for mapping the people.

1. Who are the people involved directly?

2. Are there additional relevant stakeholders? If yes, who are they?

3. What are their typical characteristics with respect to:

    ○ roles,

    ○ profession, qualification,

    ○ knowledge and skills,

    ○ experience,

    ○ context of work or use,

    ○ role in the analyzed situation or context, and

    ○ level of involvement and responsibility.

Roles: Specify the part each person played in the scenario being analyzed, such as a patient, family, or caregiver. Often, a person can play a role in a given context or event that is different from their formal profession (for example, a nurse educator stepping in to help with a resuscitation event). Describe these types of exceptions as well. Be careful to focus on the roles, not the tasks. We will get to the tasks at a later stage of the analysis.

Profession and qualifications: Specify the formal profession of the stakeholders, such as a nurse, physician, or physiotherapist. In addition, indicate if the person has acquired qualifications that are beyond the formal profession, for example an ICU nurse or an operating room nurse.

Knowledge and skills: People have knowledge and skills that often are an inherent part of their professions and qualifications. For example, an ICU nurse knows how to calculate medication dose and administer it, or is skilled in taking blood for tests. When we talk about skills, we should distinguish between technical and non-technical skills. Technical skills are part of the profession and qualifications (for example, putting in an IV). Non-technical skills have to do with various intellectual, emotional, and social skills that are associated with the technical work (for example, communicates clearly, people-friendly, and good leader).

Experience: Specify the stakeholders' experience by considering the combined impact of their education and training, certifications, and time practicing in their formal profession, and having other qualifications. Experience is often highly correlated with expertise.

Context of work or use: Specify the place where the people do their tasks. Note that later, when we go deeper into the physical and human environment, you will elaborate on this.

Level of involvement: So far, you have outlined the people involved in the event, situation, site, or procedure you are analyzing. However, ask yourself if everyone's level of involvement is the same. While all are stakeholders, they are probably not involved to the same extent. Some are more directly involved, and some are indirectly involved. For example, the ICU nurse attending to patients is probably directly involved, whereas the charge nurse or the head of the ICU is more likely to be indirectly involved.

We propose organizing the mapping information in a table as is shown in the scenario examples below. Table 6.1 shows an example of mapping people in the sepsis scenario.

Table 6.1: Mapping people in the sepsis scenario					
	Characteristics				
People	**Profession, Qualifications**	**Knowledge, Experience**	**Context**	**Role**	**Level of Involvement**
Triage RN	Nursing school (RN); Emergency Medicine training certificate	Full time; several years of experience in Emergency Medicine; variable experience in triage role	ED—triage area	Triage	Primary
Unit Clerk	Medical Unit Clerk training	Full time; years of experience	ED—triage area	Administrative	Secondary
Clinical Area RN	Nursing school (RN); Emergency Medicine training certificate	Full time; several years of experience in Emergency Medicine	ED—clinical area	Management in ED clinical area	Primary
Attending Physician	MD; Emergency Medicine training program (5 years)	Full time; several years of experience; highly skilled in evaluation and resuscitation of acutely ill patients	ED clinical area	Management in ED	Primary

If your analysis is reactive, you would probably know which people are involved. However, if your analysis is proactive, you may not know who would be involved, directly or indirectly. When you do not know who would be involved, you can characterize segments of the population relevant to your analysis, and then delineate "personas." This is a useful tool, particularly for proactive scenarios. While the actual people involved may play the role of each persona in reactive scenarios, it may still be useful to define more generalized personas for the analysis, even in reactive scenarios. One way this can help is to de-personalize the analysis, reducing threat to individuals. Another benefit is that you can include aspects that come from the experience of other people from the same population– broadening and deepening the analysis. For more about which criteria to use in segmenting the population, how many segments and personas should be constructed, and what are the useful parameters of a persona (see Pruitt and Adlin, 2010; Mulder and Yaar, 2007).

Below we will outline the key populations, segments, and persona for the sepsis scenario analysis using the detailed questionnaire.

6.1.1 WHICH POPULATIONS AND SEGMENTS ARE RELEVANT TO THE ANALYSIS

You can segment the existing or prospective user population into sub-populations. Each such segment constitutes a user profile and is uniformly characterized by these key parameters: profession, experience and specific skills, and context of work. Depending on the goals of your analysis, a segment could be nurses, for example. It could also be experienced nurses because your analysis deals with a site or a process that will only involve experienced nurses. It could even be narrower—CU experienced nurses.

6.1.2 PERSONA

You can represent each such segment by a persona. A persona is an archetype representing a specific segment in the population. It is a model representing all the people in a given population segment. A persona is not a real person, but it is a depiction of "someone" who may be perceived by us as a real person. Once we have constructed a persona, it is easier for us to proceed with the analysis dealing with a single person, as it were, rather than many people or users. It is, however, very important to emphasize that we probably have several personas in any given situation, site, or process,

In order to make the experience of doing a fit analysis on a persona similar to that of a real person, we add some details.

- **Name:** yes, give the persona a name. It will make the persona seem like a real person.

- **Picture:** having a picture of a fictional person adds another dimension of familiarity to the persona.

- **Context of work:** this parameter is a part of defining the population segment and could be added here with some more details.

- **Tasks and responsibilities:** add more details about the persona's specific tasks and responsibilities. Here you can go a bit beyond the boundaries of the specific event, situation, process, or site; so that later in the fit analysis you will have a broader understanding of what could influence the people and situation involved.

- **Typical schedule:** add information about the routine of the persona as it relates to the analysis. Consider going beyond the analysis boundaries to have a broader perspective on other possible influences.

- **Colleagues, peers, and others:** while we map the human environment later, we can start here by having some information about the contacts and work relationships of the persona with other people relevant to the analysis goals.

- **Useful tools:** outline the tools and artifacts (such as forms and other objects) used by the persona. Examine them in connection with the physical environment, outlining the location of the tools and artifacts in each environment.

- **Capabilities and limitations:** outline both intellectual and physical capabilities and limitations. These will be very useful later when you do the fit assessment and wish to include the specifics of your personas, and not only the "average" capabilities and limitations that we know of from research and human factors literature. In other words, by including intellectual and physical capabilities and limitations, the fit assessment will be more realistic, making it easier to assess.

- **Attitudes, motivations:** outline key aspects in the persona's attitudes toward the profession and role, the workplace, other people, and the organization. In addition, outline motivations as relevant to all those factors. This part of the persona will also be very useful in the fit assessment phase of the MARC process.

- **Satisfaction and pain points:** outline what the persona in happy about in the context that is analyzed, and what they are unhappy about. These aspects could be particularly useful in a proactive analysis since it would help identify needs that, when unmet, could influence performance and outcomes. They could also help suggest which needs require interventions and mitigations.

- **Favorite quote:** A final touch to making the persona more familiar and real is to add a fictional favorite quote. That quote should reflect what is truly important to the persona. It could reflect hidden needs, wishes, and aspirations. It could reflect fundamental values and beliefs, or it could uncover an important opinion. All of these can provide further support in assessing fit, recognizing emergent factors, and going on to the conclusions of the analysis.

6.2   MISSIONS, GOALS, AND TASKS

An individual, a team, and an organization all have goals. Goals could be on a very high level (for example, "*Health Canada is committed to improving the lives of all of Canada's people and to making this country's population among the healthiest in the world as measured by longevity, lifestyle and effective use of the public health care system.*") or a very specific level (for example, "*Perform a Cardio Arterial Bypass Graft surgery successfully and safely.*").

Tasks are simply what people do. Tasks are the specific actions one takes in order to achieve one's goals. Tasks can be any action—physical or cognitive. Physical tasks are actions that involve motoric and sensory-motoric actions. For example, physical tasks in a surgery are injecting, inserting, placing, cutting, pressing, moving, holding, and many more. Such tasks would be typically overt and easy to observe and capture. Cognitive tasks[1] involve perceptual, cognitive, and communication activities. Perceptual tasks may include searching, detecting, recognizing, and attending. Cognitive tasks may include storing in memory, retrieving from memory, comparing, computing, planning, solving a problem, and making a decision. Communication tasks may include asking, replying, announcing, and reading back. Later on, when we assess people's capability to perform their tasks successfully, we will revisit those task definitions.

Use the detailed checklist for mapping goals and tasks for the people involved by considering these questions.

1. What are the goals of each of the individuals and/or the team? Examples include administration, triage, diagnose, treat, and others.

2. What are the physical, overt tasks the individual or team performs? Examples include inserting an IV and administering a medication.

3. What are the physical activities required for performing the task? Examples include reaching, grabbing, walking, pointing, and inserting.

4. What are the perceptual, attentional, and cognitive tasks the individual or team performs to achieve this? Examples include monitoring relevant elements of the environment by looking, listening, and touching, noticing changes, and evaluating themeanings

---

[1]   Detects, Inspects, Observes, Reads, Receives, Scans, Surveys, Discriminates, Identifies, Locates, Categories, Calculates, Codes, Computes, Interpolates, Itemizes, Analyzes, Chooses, Compares, Estimates (from: Vincente, 1999, p. 66).

associated with these. It includes situational awareness, which reflects how each person understands what is happening with respect to the mission or task being performed.

5. What are the inter-relations between the tasks? Note, this part of the mapping is addressed in a more elaborate fashion in Section 6.2.1. Once you have the information for each of the questions, it is useful to organize it in a table.

Table 6.2 answers the above goal and task mapping questions for the sepsis management scenario.

Table 6.2: Sepsis management scenario—mapping goals and tasks			
People	**Goals and Tasks Mapping**		
	Goals	**Tasks**	**Required Capabilities**
Triage RN	Identify sepsis[2]; expedite management	Take history, take vitals, suspect sepsis, activate "sepsis protocol" (i.e., use the sepsis workbook), including communicate with MD, facilitate initial lab tests, expedite patient transfer to clinical area, communicate with clinical area RN	Be familiar with the sepsis syndrome; be able to use the sepsis workbook; communication skills, in particular be able to communicate urgency and skilled in communication techniques; multi-tasking; attention allocation/division; computer and other device literacy; work with other people; ability to adapt and change actions and directions; manage pressure and urgency, time pressures
Clerk	Register the patient	Register patient in the electronic system; identify an unwell patient; communicate concern to triage nurse	Basic clinical skills, communication skills—assess urgency
Clinical Area RN	Adequately provide all elements of sepsis care	Establish IV access; draw blood for tests; administer fluids and medications; monitor patient; communicate concerns to MD	Technical skills (IV access, infusion, etc.); other clinical skills (identifying an unwell patient); communication skills (urgency, closed loop); multi-tasking; attention allocation/division; computer and other device literacy; leadership; work with other people; ability to adapt and change actions and directions; manage pressure and urgency, time pressures
MD	Identify/diagnose sepsis; adequately manage a septic patient	Take history and perform focused physical exam; order appropriate initial fluid resuscitation, lab and other tests, and antibiotics; monitor response to treatment and respond appropriately; insert a central venous catheter if needed	Be familiar with the sepsis syndrome; skilled in resuscitation, including technical skills (central line insertion); communication skills—perceive urgency from RN; multi-tasking; attention allocation/division; computer and other devices literacy; leadership; work with other people; ability to adapt and change actions and directions; manage pressure and urgency, time pressures

---

[2] It should be noted that the overall goals of the triage nurse is to triage. For the sake of the example brevity, we focus on the sepsis aspect of the scenario.

6.2.1   TASK ANALYSIS: MAP THE INTER-RELATIONS BETWEEN THE TASKS

A task analysis is a breakdown of someone's goals into the major tasks and sub-tasks (that is, into actions, including cognitive actions) performed in order to achieve the goals. "Someone" could be an individual, team, or organization. There are various approaches and techniques to conduct a task analysis. The Hierarchical Task Analysis (HTA) is the most common form. It uses a hierarchical structure to represent the goals and tasks (see Figure 6.2, below). Task analysis could be descriptive, modeling the present state of tasks. This would be a step-by-step representation of how the task is actually being performed. Task analysis could be normative, modeling the desired state of tasks. This describes how the tasks should be performed, rather than how it is actually being performed. Hierarchical task analysis (HTA) involves the decomposition of each user goal into tasks and actions, often depicted as a basic task structure or mapping diagram.

Figure 6.2: A generic example of a hierarchical task structure.

The following task analysis in Figure 6.3 is a normative example when sepsis may be involved, that is, it describes how sepsis should be managed:

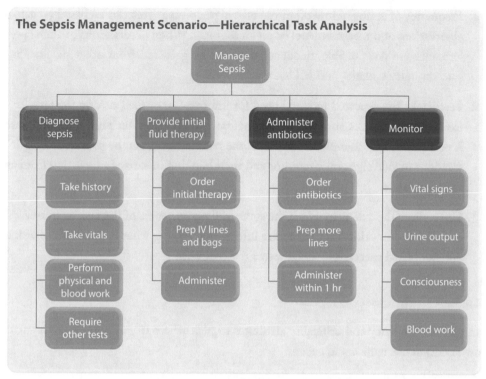

Figure 6.3: The sepsis management scenario—hierarchical task analysis.

Beyond the basic task structure mapping, it is sometimes important to analyze the tasks with respect to various parameters, such as how often the task is performed and how difficult it is to do. This analysis is usually performed in a tabular form. Here we make a distinction between the fully detailed task analysis and a short task analysis of what the user does or is supposed to do. A full, detailed task analysis could use many parameters.[3] The short task analysis involves only three analysis parameters. These three primary parameters are sufficient to analyze the task and derive significant implications for many purposes, such as interaction and usability design. The parameters are as follows.

---

[3] Task trigger, duration and variability, physical and mental activities, frequency, complexity and difficulty, contextual factors, information required, tools required, feedback, criticality, and outcomes.

1. **Frequency of action:** How often the task is performed. This can be based on actual observations and measurements or on an estimate. When observations of actual performance are not possible, frequency data can be gathered from other stakeholders (such as, nurses, clerks, and MDs).

2. **Feedback:** The expected required need for information or feedback on the task. This feedback can be cues and information before, during, and after performing the task is performed. For example, feedback on the patient status can be provided using indications of vital signs on a monitor and their blinking in case of a deviation from an expected range.

3. **Criticality:** The extent to which the successful performance of the task or sub-task is essential and/or critical for the entire interaction. Asking what happens if the task is not completed successfully helps assess the criticality.

Task Analysis Implications for Design

You can use the frequency and criticality attributes to plan where to put options in the workflow and how to design mechanisms to ensure:

- ease of interaction (for example, for complex or rare tasks),

- efficiency (for frequent tasks the user will have lots of practice with), and

- protection from errors (for critical tasks).

Feedback refers to the amount, content, and location of information given to the user. The straightforward implications are that frequent actions should be more immediately operable, and tasks that require more feedback get more screen "real-estate" and/or longer exposure.

Using these considerations, a unique added value of the analysis is in rapidly and clearly identifying the trade-offs between these parameters that can have a large impact on user interface architecture and interaction design. For example, the analysis may show that in a Hospital Information System there are some low-frequency tasks such as recovery and restore of data after a major power failure yet with a high criticality value that requires detailed information to be displayed on what was lost, what needs restoring, the progress of the recovery process, and its outcomes.

Task Analysis Implications for Usability Testing

You can use the outcomes of the short task analysis to determine the objectives of a usability test and then focus on the relevant tasks. The task analysis can be used to generate questions such as the following.

- Should we focus on the critical tasks or the low frequency yet highly critical tasks?

- Are the frequently used tasks quick to find and use?

- Are the protection mechanisms for critical tasks effective?

In addition, you can use the task analysis to design test tasks and scenarios based on the outcomes of the analysis. For example, looking at the task analysis, you can ensure that a frequent task is represented as such in the test scenario. If there is a critical task, the scenario should include appropriate consequences.

6.2.2   TASK ANALYSIS IMPLICATIONS TO SEVERITY RATINGS

Frequency and criticality are important parameters for evaluating the severity of a lack of fit between people's capability and the demands of a task. The severity of a failure is a function of its likelihood to happen and its impact. The more frequently a person performs a task they are not capable of doing, the more likely the failed task will occur. The more critical the task, the greater its impact.

Note that after the mapping phase, when you assess the fit between the requirements of the task and the human capabilities and limitations, you will need to determine the severity of issues. The frequency and criticality parameters that you have addressed here as part of the task analysis can contribute to that analysis.

6.2.3   TASK FLOW MAPPING

It is important to consider the task flow: the sequence by which people perform their tasks. This sequence could be linear, conditional upon various contingencies, or proceed in several inter-related and parallel sequences (particularly when examining teamwork, but also when multi-tasking). The simplest way to model task flow is by a flow chart which shows the start and end points, the progression of tasks, and the conditions and dependencies between them. However, if you wish to model a more complex task flow involving several stake-holders and/or interaction with devices, then an appropriate way to show it is by using a "swim lanes" diagram. Such a diagram (see Figure 6.4) has a "lane" (shown as a column) for each participant (whether participant is a person or a device) in the task flow. Each action is inserted into the relevant "lane," and any interactions or task flows between the various participants are shown by links between them in the diagram. The other dimension (the vertical one in the example below) shows time progression.

Figure 6.5 shows a swim lanes diagram for the sepsis management scenario.

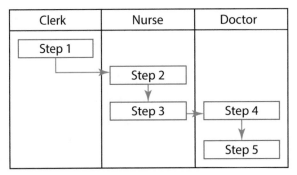

Figure 6.4: **Example for modeling task flow using the swim-lanes visualization**

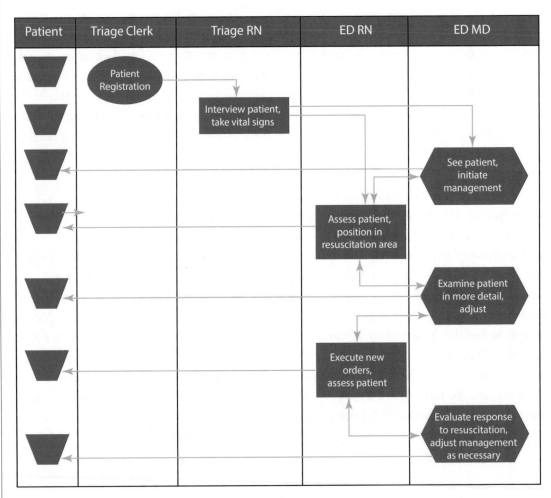

Figure 6.5: Swim lanes diagram for the sepsis management scenario.

6.3   PHYSICAL ENVIRONMENT

When we first consider the physical space we are often tempted to say things such as "there is not enough space," "it is too crowded," or "it is too far to reach from one place to another when I have to." These are examples of great intuitive assessments of the fit between space and layout to the demands of people and their tasks. However, before we begin to assess, we must first have the facts. We must first map the physical space available to individuals or groups/teams to perform their tasks.

In mapping the physical environment, the factors to map are:

- space and layout;

- artefact and device locations and access;

- layout and locations—link analysis;

- ambient conditions; and

- human-machine interface.

There are a number of techniques you can use for mapping. Two recommended approaches are as follows:

- Using the detailed checklist—this involves answering the questions as accurately and completely as is relevant for your goals.

- Using link analysis—see the next section for more information on how to perform it.

6.3.1   SPACE AND LAYOUT

Spatial arrangement typically influences task flow: the sequencing of actions people do to achieve goals. You can analyze the space and layout using the detailed checklist and using link analysis.

Analyzing Space and Layout Using the Detailed Checklist

Answer the following questions to use the detailed checklist for mapping the space and layout.

- What space is available?

- Is the space divided into sub-spaces? If yes, what are the criteria for this division?

- How are the sub-spaces arranged spatially?

- Are there access paths and routes?

- Where are the entry and exit points?

- Where are the locations of doors and windows or any other fixed elements?

Here are answers for the sepsis management scenario. The link analysis diagram follows.

- The Emergency Department (ED) comprises a large waiting area and several clinical sections where care is provided. The registration and triage processes take place in the waiting area. A patient arriving independently to the ED would first register with the clerk, and then approach the adjacent Triage desk to be seen by a nurse. If no emergent condition is identified, the patient will likely be instructed to take a seat and wait to be called in. There are about 20–25 seats available in the waiting area, a hall of roughly 1,000 ft^2.

- A large door separates the triage from the clinical areas.

- The clinical management is provided in 4-5 sections (A–E), arranged in a circle around a common corridor, and each contains about 10-12 stretchers. "Section A," the "high acuity" area, is the closest to the main entrance and triage area. It contains three "Resuscitation Rooms," where equipment necessary for various life-sustaining therapies is stocked routinely. The Charge Nurse is usually present in Section A. Typically the nursing stations are located at the center of the different areas, surrounded by the patients' stretchers. There are multiple rooms and cabinets with medical equipment, supplies and medications, several washrooms, and working stations for clinicians. There is an X-Ray room at about the center of the ED, however, CT and other Radiology services are located outside of the ED.

- The ED waiting area was quite full of people, patients, and healthcare workers.

Using Link Analysis to Analyze the Physical Environment

Link analysis supports mapping, design, and evaluation of the layout of any elements. It could be the layout of a kitchen, control room, control panel, operating room, medication room, or equipment or computer display elements (e.g., a whiteboard or a large computer display). The technique maps the relations and interactions between all possible system components. The mapping can be done as a graphic diagram or as a link matrix table. We will focus here on the graphic diagram technique because it is powerful for both performing and communicating the analysis.

Mapping the space and layout literally refers to drawing the spatial arrangement of all elements in the workspace. This should include other rooms and routes between the rooms (access paths) that are relevant to the task flow.

The space and layout is often illustrated using a link diagram or a spaghetti diagram. Here are examples for each.

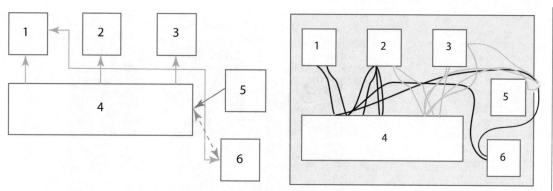

Figure 6.6: Example of a link diagram (left) and a spaghetti diagram (right).

The link analysis technique has several steps.

1. Decide on the elements to be mapped (rooms, hallways, devices, artifacts, people, information items, actions, or all).

2. Define the nature of the links (frequency of the association between the elements, operational sequence or flow, importance of the associations between the elements, etc.).

3. Map the links, in a link or a spaghetti diagram.

4. Analyze and interpret the links in terms of element locations and their proximity with respect to tasks and interactions.

Link analysis can be used for both reactive and proactive scenarios. In reactive scenarios, use it to assess an existing layout. In proactive scenarios, use it to express the requirement of how a space should be laid out and where the people and artifacts should be located. The link analysis can provide guidance in terms of:

• workflow and transitions: to delineate the existing or required transitions between tasks and artifacts at various locations (for example, mapping the routes people take in a given space from one location to another when performing a task);

• grouping and clustering: to delineate the existing or required commonalities among tasks and/or artifacts that can be grouped into physically proximal units (e.g., group all frequently used medications in physical proximity to support the frequent use); and

• space layout: to provide the basis for space organization and element arrangement according to the links and associations between them (e.g., place devices close to the location of people interacting with them).

Here is a spaghetti diagram based on the sepsis management scenario.

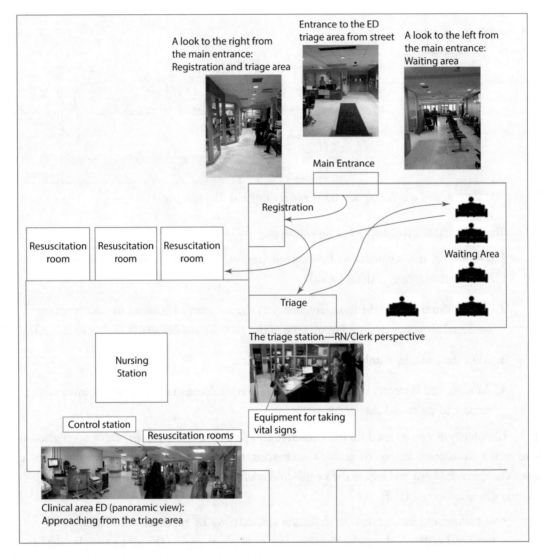

Figure 6.7: Spaghetti diagram based on the sepsis management scenario.

6.3.2   ARTIFACT AND DEVICE LOCATIONS AND ACCESS

On top of the overall space and layout representation, map the spatial arrangement and location of artifacts and devices used to perform tasks and achieve goals. Artifacts include any electronic and non-electronic devices or aids, tools, substances, clothing, etc.

Mapping Artifact and Device Locations and Access Using the Detailed Checklist

Answer the following questions to use the detailed checklist for tabulating locations and layout.

- Where are all the required elements (people, displays, anything) located?

- Where are the artifacts used by individuals to perform tasks located?

- Where are the artifacts required by several people to perform their tasks located?

- Where are the people in the space located relative to the location of artifacts, devices, and information they require to perform their tasks?

Here are answers to these questions for the sepsis management scenario.

1. Triage desk for taking vital signs, computer documents for history-taking; phone; forms; same equipment in the Emergency Department (ED) itself if patient taken directly there; used by nurse and/or physician; in triage all on desk next to nurse; in ED, computer.

2. Housekeeping big cart was located just at the center, partially blocking the way.

3. Artifacts: sepsis workbook, equipment for vital sign measures (BP, temp, O_2 saturation), monitor, fluid bags, lines, drugs, O_2 supplementation devices, Foley catheters.

4. Various pieces of equipment were left all over the place. Some were still dirty after being used.

Mapping Artifact and Device Locations and Access Using Pictures and Diagrams

Figure 6.8 shows a visual presentation of artifact locations in the sepsis management scenario.

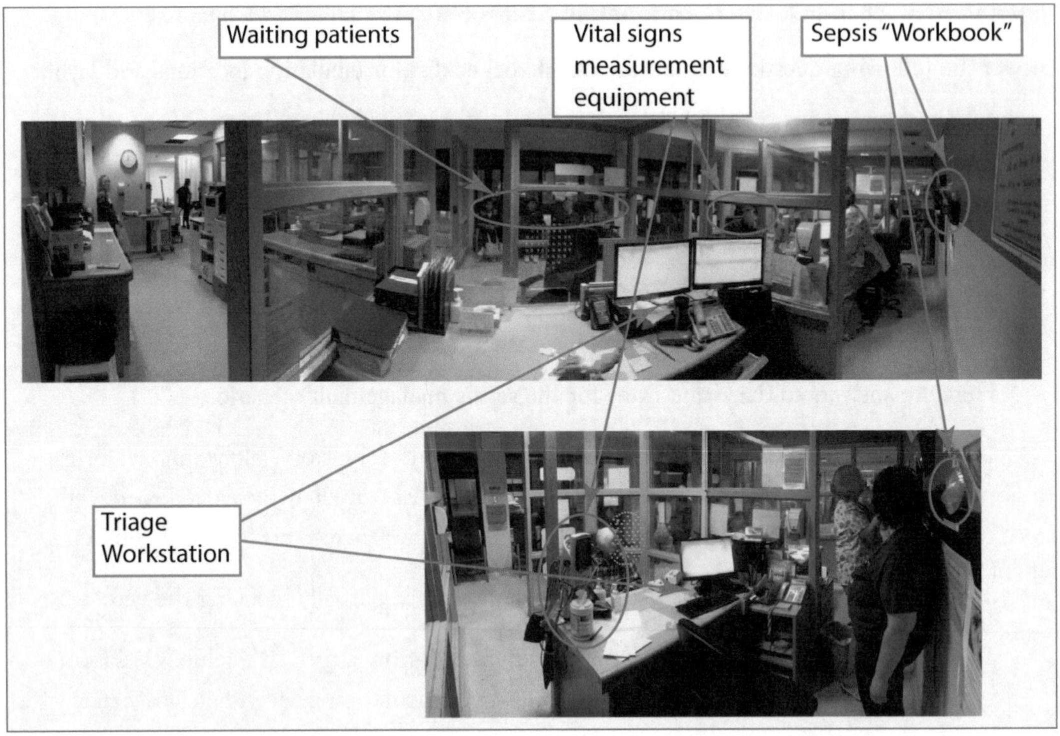

Figure 6.8: Visual presentation of artifact locations in the sepsis management scenario.

6.3.3   AMBIENT CONDITIONS

The ambient conditions refer to lighting, noise, temperature, odors, and vibrations. Later, when you perform the fit assessment, you will use OSHA standards that provide the appropriate levels required for each parameter. Deviation from the recommended levels can result in physical, cognitive, and emotional disruption or even harm.

Mapping Ambient Conditions using Measurement Devices

You can use numerical measurements of light, noise, and temperature to map ambient conditions. Even smart phone apps can be used for taking measurements. For example, this simple illuminance meter measures the amount of light that reaches the human eye.

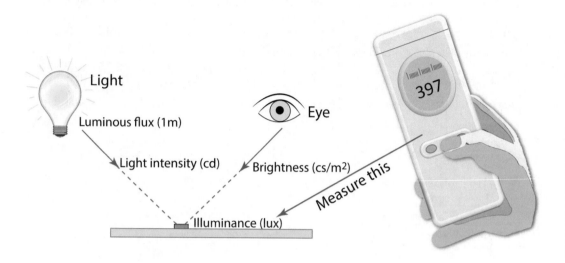

Figure 6.9: Measuring ambient light with an illuminance meter.

Mapping Ambient Conditions Using the Detailed Checklist

You can map ambient conditions by answering the following questions in the detailed checklist for mapping the ambient conditions.

1. What light is available to perform the task? What are the light levels? Are there reflections anywhere in the workspace?

2. Is there noise? What is the nature of the noise? What is the noise level?

3. What is the temperature in the work environment?

4. What is the air quality in the work environment?

5. Are there vibrations in the work environment? If yes, does the frequency or intensity of vibrations create disturbance?

6. Are there any usual or unusual odors in the work environment?

The following are answers to the ambient condition questions for the sepsis management.

1. Fluorescent light is used throughout the triage area and the main clinical area; little daylight penetrates the triage and waiting area.

2. Noise can reach high levels. The main source of noise is the usual commotion such as people talking, phones ringing, pagers beeping, some medical devices such as monitors and infusion pumps, and other office machinery such as printers.

3. Temperature is controlled through air conditioning throughout the ED.

4. Air quality is good.

5. No vibrations exist.

6. Odors are not common, but may include human scents—natural and artificial,.

6.3.4   HUMAN-MACHINE INTERFACE

The tools and devices people work with are a very important part of the physical environment. In many cases, these tools and devices are interactive, electronic, and software-based. Interactive technology plays an increasingly decisive role in what happens in a given situation or site. Consequently, the mapping of the Human-Machine Interface (HMI) as a critical part of the devices people work with is very important to our mapping. Taking the metaphor of a conversation between people and the interactive technology, the HMI is typically composed of three fundamental elements (see also Figure 6.10).

1. **Machine Display:** the part of the system that "talks" to users. The term "display" is used here in a very broad sense: it is *anything in the system* that provides users with information and feedback. That *anything in the system* could be the part conveying information in any of the human modalities: visual, auditory, haptic and tactile, and olfactory. It could be something that is part of the familiar computer setup, such as a monitor or speakers or a vibrating mouse. However, it could also be ambient and ubiquitous (for example, a public announcement system, large displays) in the environment.

2. **Human Control:** the part that enables the human user to "talk" to the system. That part refers to anything in the interface that allows the user to control and execute operations. It could be something that is part of the familiar computer context such as the keyboard or mouse, or even touch-based interactions. However, it could also

be voice command, eye-gaze control, gesture and movement recognition, or even brain-computer interface inputs that are translated into commands.

3. **Interface:** where the "dialogue" between the human user and the machine aimed at starting and successfully completing the task occurs. The details of the dialogue include the way the dialogue starts, how it proceeds, and how it is completed or aborted. The dialogue typically consists of steps and phases required to start and complete tasks. In the field of Human-Computer Interaction, we also talk about interaction styles such as: typing in commands, filling in forms, selecting from menus, directly manipulating elements on the screen, and finally other natural interactions such as gestures, speech, gaze, and thought.

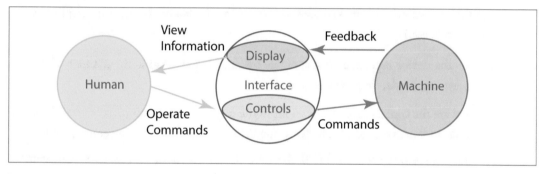

Figure 6.10: A schematic representation of the Human-Machine-Interface (HMI) components and flow.

The appropriate design of the HMI supports workflow and tasks, facilitates subjective experience, and can prevent or at least minimize human error.

6.3.5   MAPPING HUMAN-MACHINE INTERFACE USING THE DETAILED CHECKLIST

Answer the following questions to use the detailed checklist for mapping the HMI in the physical environment.

Display

1. What is the subject matter (content) of the information? (For example, vital signs, HR, BP, etc.)

2. What are the information characteristics? (For example, high-low, exact number, trend, etc.)

3. What is the level of detail in the information? (For example, do we need to show very high detail or is it enough to give ranges? Is a very accurate task required or is qualitative information enough?)

4. What is the rate of update of the information?

5. What is the level of tolerance for errors in the information?

6. In what ways or modalities is the information conveyed?

7. In how many ways is the information conveyed?

8. How is the information presented and conveyed? (For example, visually: text vs. graphics, numeric vs. analog, size, color; auditorily: sound vs. speech, amplitude, frequency, etc.)

9. Does the device include a haptic, tactile display? If yes, what is the manner of the display: amplitude, frequency, temperature?

10. What are the contextual considerations in the way the information is conveyed? (For example, indoors vs. outdoors, lighting conditions and reflections, movement, etc.)

The following are answers to the HMI checklist display questions for the sepsis management scenario.

1. Usual electronic medical record content.

2. Demographic and clinical information.

3. Depending on the information type, some is quite detailed, other may be general content.

4. After registration, information may be updated as frequently as new information such as vital signs, assessment and test results is generated.

5. Manually entering erroneous information (by human operators) is possible. Intercepting an error is up to the individual. Test results are mirrored in the system as reported by the lab.

6. Information may be entered manually by different operators or transmitted electronically from different users such as the lab and the imaging department.

7. At least two ways—manual entering and electronic transfer.

8. Information is presented on a computer screen.

9. N/A

10. To enter or review information about a patient, the provider has to log on to the system and to open up the specific patient's electronic chart.

Controls

1. What is the controlled function or action? (For example, raise or lower threshold, etc.)

2. What is the human modality for the control? (For example, hand, finger, foot, speech, etc.)

3. What are the control actions? (For example, Push, pull, press, tap, rotate, speak, look, move, think, etc.)

The following are answers to the HMI checklist questions for the sepsis management scenario.

1. Entering and/or accessing information.

2. Fingers and hand.

3. Keyboard and mouse.

Dialogue

1. How do users start the dialogue, the task?

2. Is the dialogue continuous or in discreet steps?

3. Can the user undo or reverse steps?

4. What is the style of the dialogue? (For example, selecting commands from menus or keying numbers based on a voice menu? Direct manipulation?)

5. How do users complete or abort the dialogue?

The following are answers to the HMI checklist questions for the sepsis management scenario.

1. By logging on to the system and to the specific patient's chart.

2. The dialogue is discreet. Users typically review and/or enter information repeatedly over the process of managing the patient.

3. Once information is entered the user cannot undo it. There is an option to add a comment regarding erroneous information.

4. Information can be accessed and observed as per the user's will. Users can enter information using mostly free text and sometimes selecting from a menu.

5. The dialogue is completed by saving an entry and logging off. Not uncommonly a user would abort the dialogue through just walking away from the workstation.

6.4   HUMAN ENVIRONMENT

We started our mapping with people (see Section 6.1). There we focused on the main stakeholders. Now it is time to expand the mapping of people to a broader scope and cover additional aspects that could be influential. The human environment is composed of several levels. An individual is typically a member of a group or a team. The family of a patient can be considered as part of the human environment. A team, or even the individual, can be part of an organization. In healthcare, an organization could be a unit in the hospital or the hospital itself. Individuals, teams, and organizations would typically be part of national frameworks such as national or professional associations, the province, and the entire country. We could even consider international frameworks, such as the World Health Organization.

Groups and Teams: We began by mapping the key stakeholders involved in the case or site or process. Now we move on to map the entire human context of those stakeholders. Several co-located people engaged in activities that are linked could be considered a group, even when the people are distributed in different locations. We tend to distinguish between teams and groups. Teams are most often defined as "a distinguishable set of two or more people who interact dynamically, interdependently, and adaptively toward a common and valued goal/object/mission, who have each been assigned specific roles or functions to perform, and who have a limited life span of membership". Teamwork is not only characteristic of many healthcare contexts, but is very often an inherent aspect of the care process, and is critical to patient safety because healthcare workers interact with each other in order to achieve a common goal successfully and safely. To accomplish the common goal, all members of the team must perform their roles and tasks with full and continuous comprehension and awareness of the dynamic situation. To do this, they rely on continuous coordination, communication, and information sharing with their teammates. Effective teamwork is often a function of its structure; a team can have a leader managing operations and a workflow with a clear "chain of command" (a hierarchical structure), or the team can operate without any central management (a flat or democratic structure).

Answer the following questions to use the detailed checklist for mapping the broader scope of relevant people.

1. Who are the other people in the workspace (healthcare professionals, patients and families, any other personnel)?

2. What do they do?

3. Are they members of a group or a team with these other people? Are the key stakeholders members of a group or a team with these other people?

4. What is the structure of the team (flat vs. hierarchical)?

5. Is there leadership? Who is/are the leaders? How did they become leaders?

6. What do team members know about each other?

7. How do team members communicate? How often? What about?

8. How do team members coordinate and collaborate?

9. What is the definition of roles?

10. What is the allocation of workload?

11. What are the expectations and understanding of everyone's role and what is supposed to be done and when (mental models)?

Here is an example of aspects of groups and teams in the sepsis management scenario.

1. In the triage area: patients and care providers, registration staff, triage nurses, security staff, volunteers.

2. Patients and escort: waiting to be seen; talking—over the phone, or among themselves; staff members: attend to their respective roles in the area, including charting, interviewing, giving instructions, etc.

3. Stakeholders compose of individuals with ad-hoc teams. Typically, a single registration clerk or two; one or two triage nurses; multiple ED nurses and physicians, including attending and resident physicians at different levels.

4. No real hierarchy among nurses and registration clerks; no hierarchy among triage and clinical area nurses; doctors in a position to give orders, make decisions, and thus perceived hierarchical gradient between doctors and nurses.

5. Triage nurses, out of the nature of their activities (e.g. assigning illness severity to patients, prioritizing care) are the natural leaders in the triage area. Yet, leadership role may not be formalized and accepted by all.

6. Providers usually know each other, at least within profession (i.e., clerks, nurses, doctors). There may be newer providers that may not know others/vice versa. Obviously, providers usually do not know the patients prior to their admission.

7. Communication is usually ad hoc, and verbal, face to face. Phone is also an option.

8. Coordination and collaboration usually occurs verbally, on the fly.

9. Roles are clearly defined as per the profession (e.g., clerks, nurses, doctors)

10. Workload varies and is for the most part unpredictable. Some hours in the day may be quiet and others very busy. Similarly, some particular events such as acute stroke protocol and trauma team activation may suddenly and substantially increase workload.

11. For the most part, providers have a good understanding of their own roles and their colleagues' roles. In the case of sepsis protocol, once the protocol is initiated, providers usually have a good understanding of their roles.

Organizations and organizational culture: An organization is an entity incorporating people, tools, and devices. All the members of an organization share a goal common. Organizations have structures that refer to the arrangement of roles and positions, authority, responsibilities, and the inter-relations between all of these. An organization's culture is defined by its shared beliefs and attitudes, values, and behavioral norms.

Use the detailed checklist.

1. What is the division and definition of roles within the organization?

2. What is the organizational structure?

3. What are the staffing practices and policies?

4. What are the organization's approaches toward workers, safety, quality, and accountability?

5. What kind of learning takes place in the organization?

6. How is information transferred within the organization?

7. What are the development, training, and certification opportunities?

Here is an example of aspects of organizations and organizational culture in the sepsis management scenario.

1. There are role definitions and responsibilities.

2. Nurses and clerks are hospital employees. Doctors are not; however, they are granted privileges to operate within the hospital, and they are to follow hospital policies.

3. Sepsis protocol often is not followed. In the triage area there will be one or two clerks at a given time, and one or two nurses. There are no doctors in the triage area. The ED's main clinical area is staffed by clerks, nurses, doctors, and other staff members.

4. Employees are expected to follow hospital policies; they are held accountable for their actions.

5. Learning takes place ad hoc, especially when a new device or process is introduced. Time constraints and other considerations preclude effective learning quite often, and thus staff would often learn on the fly.

6. Email is commonly used. In-person interaction is less common.

7. Development and further training is for the most part to the discretion of the employee (in the case of nurses and clerks).

Rules, regulations, and policies: These are the documentation of an organization's missions and visions, goals and objectives, allocations of roles, scope of authority and responsibilities, prescribed procedures, desired behaviors and responses, and overall guidance on "how to do things."

Use the detailed checklist.

1. What are the available policies?

2. How are the policies designed and communicated?

3. Are there too many workarounds? If yes, what is the reason?

4. How is workload allocated and managed? Consider work hours, shifts, rotations, rest periods, night work.

5. What is the practice in terms of sticking to rules and policies?

Here is an example of aspects of rules, regulations, and policies in the sepsis management scenario.

1. The sepsis protocol is available as an electronic file within the hospital's intranet, and also in a physical binder in different locations in the ED.

2. The sepsis protocol was designed by a multi-disciplinary team, including ED providers.

3. Teaching sessions were held for most doctors and about two thirds of the nurses. It was almost communicated often over email. A survey of ED providers revealed that the protocol is well received and easy to use. However, often users forget to use it, and they can't find it.

4. Nurses work 12-hr shifts. The triage position is often rotated throughout the shift. Breaks are scheduled into the work day. Doctors work 8-hr shifts.

5. Providers are generally disciplined and respect rules and policies. However, the prevailing rules may represent culture more than written policy, which providers usually are unaware of.

6.5   INTERIM BRIEF #1

Now that we have mapped out the lower tier of the pyramid, it is time to pause and review what we have so far. We have completed the mapping phase of the MARC methodology. At this point we should check ourselves before proceeding to the fit assessment phase of the methodology. We propose doing that by preparing a brief using the following template.

Brief Goal: The purpose of this fit assessment document is to validate and evaluate the mapping of the contextual factors.

Brief Objectives: The purpose of this document is to:

• summarize the mapping;

• present the findings to stakeholders;

• validate the correctness of the mapping and its relevance to the objectives of the analysis.

Brief Requirements: Based on your observations and interviews, summarize and present the mapping of users, tasks, and environment—both physical and human. The brief should answer the following.

1. Who are the people involved?

2. What do they do?

3. How do they do it?

4. What do they use?

5. Where physically?

6. Where organizationally?

Your brief should, as a minimum, include information about:

1. Background: Describe the context, for example, in terms of the site, situation, or event.

2. People: List the people involved with a brief description of each one's role.

3. Tasks: List the tasks per person. Only include the key tasks.

4. Physical environment: Describe the physical environment in words and/or a sketch (or photo if you could take one).

5. Devices and tools: List the tools and devices that play a role in the case. They can be low-tech artifacts such as paper pads or white boards, or high-tech such as computers and monitors, smart pumps, etc.

6. Human environment: Specify the unit in the hospital, the specific sub-unit if relevant, the organizational structure within the unit, and any formal protocols and procedures they follow.

7. Additional points: Include any other points you noted that are directly relevant to the mapping of people, tasks, and environment.

Consider summarizing the key mapping elements in Table 6.3.

Table 6.3: Key mapping elements

Unit of Analysis	Factors	Sub-factors	Key Mapped Factors
	People		
	Goals and Tasks		
	Physical Environment	Physical space	
		Layout of sub-spaces and general locations	
		Artifact and device locations and access	
		Ambient conditions	
		HMI of devices and tools	
	Human Environment	Teams and groups	
		Organizations and organizational culture	
		Rules, regulations, and policies	

Here is the key mapping elements table, filled in with the sepsis management scenario.

Table 6.4: Key mapping elements—the sepsis management scenario

Unit of Analysis	Factors	Sub-factors	Key Mapped Factors
	People		Physicians, nurses, clerks
	Goals and Tasks		Goals and Tasks Goals: Providing adequate management for septic patients. Main tasks: Diagnose sepsis, provide initial fluid therapy, administer antibiotics, monitor

		Physical space	Triage area followed by direct access and passage to the ED's main clinical area
	Physical Environment	Layout of sub-spaces and general locations	Busy waiting and triage area; separate from the main clinical area; space for patients to sit down (disappear?)
		Artifact and device locations and access	Face-to-face interaction with patient; vital sign measurement equipment; sepsis workbook (protocol) in a binder at the side and behind
		Ambient conditions	Noise; fluorescent lights; air conditioned
		HMI of devices and tools	Usual computer workstations; electronic health record, info entered by different people at different places
	Human Environment	Teams and groups	Skilled providers; roles defined; yet, team structure, including leadership roles, not defined
		Organizations and organizational culture	Sepsis protocol often not followed; difficulty allocating time/resources for education; difficulty maintaining continuous learning
		Rules, regulations, and policies	Written policy usually not followed. Strong culture of good patient care

CHAPTER 7

Assess Fit

In Chapter 6, we mapped the foundation tier of the conceptual framework. That is, we collected information about the context consisting of people, tasks, and environment. By sharing the interim brief and getting feedback from stakeholders, we validated that mapping.

Now it is time to do the assessment.

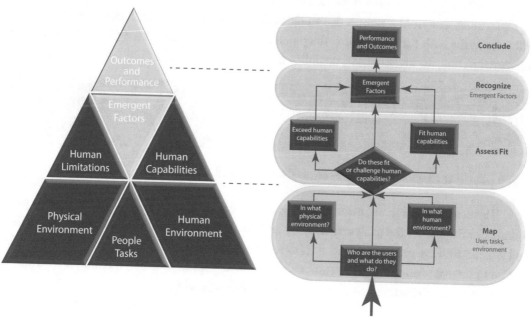

Figure 7.1: Assess the fit.

In this phase of the methodology, you assess *the extent of fit* between the "demands" of the task and the support (or lack of support) provided by the environment, on the one hand, and people's capabilities to meet those demands, on the other.

The capabilities and limitations of people are critical. They moderate how people perform the tasks in a given environment and whether or not they achieve their goals and desired outcomes.

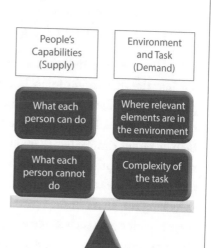

For whom do we assess the fit? Assess only the relevant factors. What's relevant? Recall earlier on in the process we determined the goals and objectives of the analysis, and one of the decisions we made was about the unit of analysis—the individuals, teams, organizational units, organization, site, and/or system. The fit you assess now is with respect to the unit(s) of analysis you set as within scope.

What does it mean "the extent of the fit"? We assess the extent to which the demands of the tasks are within the capability, at the edge of capability, or exceeding the capability of each unit of analysis. This assessment can vary in its formality. When assessed less formally, it can be a completely intuitive or well-informed guess. For a more formal assessment, you can use research, for example, using data from a prospective study in the field or in a simulator. Since the extent of fit will serve as the basis for determining the problems their severity, and the significance of the most influential factors, we suggest using the following "Extent of Fit Scale."

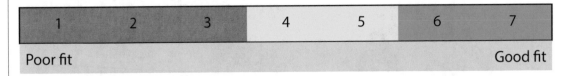

Figure 7.2: Using a continuous scale to assess the extent of fit.

This scale is continuous, in contrast to giving a very categorical rating such as poor vs. good. A scale like this allows the analyst to identify and distinguish between various levels of poor fit as well as good fit. Remember, the good fit is as important as the poor fit in this analysis. Eventually, we aim to intervene and mitigate factors associated with poor fit, but at the same time we aim to preserve and even strengthen the factors associated with a good fit.

7.1   A BRIEF OVERVIEW OF SOME RELEVANT HUMAN CAPABILITIES AND LIMITATIONS

Before we overview the relevant human capabilities, imagine a simple scenario: a nurse is walking into the medication room in a post-surgery unit in order to get a medication for a patient, prepare it, and then go back to the bedside to administer the medication. The nurse enters the medication room, looks for the medication, finds it, walks over, reaches and takes the medication. Sounds rather simple, but there are several processes that underlie this very simple scenario that can help us identify various human capabilities (and limitations) involved in such a scenario. In order to find the medication, the nurse searches visually, detects, and identifies the medication she is looking for. These involve perceptual and attentional capabilities. In addition, they may involve cognitive processes such as retrieving the name of the medication from memory or other names of the same medication in order for the nurse decide which medication she should pick from what she sees. These involve cognitive capabilities. Then she has to walk over and reach or bend, to take the medication. These involve motor actions that result from the conclusion of the perceptual and cognitive processes.

We can model the processes in this simple scenario with the following typical human perception, cognition, and action model (see Figure 7.3).

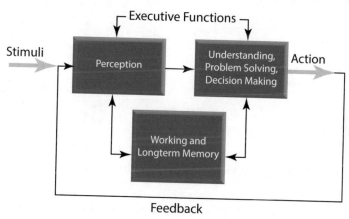

Figure 7.3: A human information processing and action model (adapted and simplified from Wickens et al., 2015).

The medication room with everything in it provides the nurse with many *stimuli*. She starts a *perceptual* process whereby she visually searches the room, maybe *attending* directly to where she expects to find or knows she will find the medication. This perceptual process of detecting and identifying the medication is supported by *memory*. Once found, she makes the *decision* that this is the medication she will take. Then she executes some *action* to reach and take the medication. The *feedback* of this action may generate new stimuli in her environment, such as noticing other relevant medications, noticing that this is the last one, or noticing that it has expired.

We use a human information processing and action model to guide our brief review of the human capabilities and limitations that are most relevant to our human factors analysis. For a more comprehensive coverage of human capabilities and limitations, consult sources such as Wickens et al. (2015) *Engineering Psychology and Human Performance*, Wickens et al. (2013) *Introduction to Human Factors Engineering*, Sanders and McCormick (1993) *Human Factors in Engineering and Design*, and the classic compendium on human performance, Boff and Lincoln (1988) *Engineering Data Compendium: Human Perception and Performance*.

Perception: The process and modalities (senses) with which one attends to and receives information, including all the factors that influence the adequate reception of that information. Visual and auditory perception are the primary modalities considered here.

Visual perception: the reception of information through the eyes. The key factors to consider are as follows:

- **The Field of View (FOV):** The horizontal and vertical range visible to the eye. We focus on elements in the center of the visual field and tend to see fuzzy in the periphery of the field.

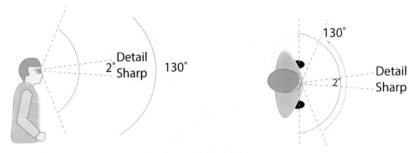

Figure 7.4: The parameters of a typical human Field of View (FOV).

- **Acuity:** The ability to discern details in the visual field. The area within which one can focus on details such as text is rather small, about 2-5° in the visual field. Focusing on small details out of that range requires eye movements and sometimes head movements. However, as you can see in Figure 7.5, there are various elements in the visual scene that can be recognized and identified even farther from the gaze direction.

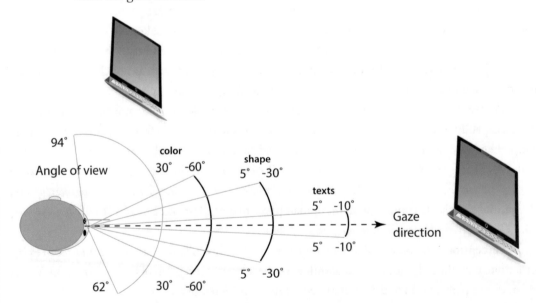

Figure 7.5: The typical elements that can be recognized and identified in different places in the FOV.

○ **Lighting conditions:** The influence of too much light or too little light on perception. High brightness and reflections can interfere with proper perception. In addition, vision deteriorates with low light levels. With low lighting, it is harder to discern details and perceive colors.

○ **Color perception:** Different wavelengths within the visible electromagnetic spectrum are perceived and interpreted as colors. We perceive three dimensions in colors: the color itself, its saturation or purity, and the level of brightness. Remember, up to 10% of the male population has some form of color blindness. In addition, ambient lighting can influence color perception.

○ **Visual search:** The process by which one visually searches for the desired elements in the visual field. The number of elements in the scene, their size, their colors, their location, and the level of similarity among them are factors that influence the accuracy and speed with which we find what we are looking for.

Auditory perception: The perception of information through the ears. The key factors to consider:

○ **Auditory perception is multi-directional:** One can hear and receive information from any direction.

○ **Localization:** We are capable of identifying the direction from where a sound comes.

○ **Noise:** Auditory noise influences the ability to perceive and identify sounds and speech.

Attention: The ability to concentrate on and receive a specific stream of information, regardless of the modality:

○ **Focused attention:** Focusing on a specific element and ignoring anything else. A negative consequence of focused attention is inattention blindness—the phenomenon whereby one is blind to what may take place in the environment and situation.

○ **Selective attention:** Being primed to attend to various elements that are relevant or important at a given time. A negative consequence of selective attention is change blindness—the phenomenon of being blind to changes in the environment or situation.

○ **Divided attention:** Focusing on several elements simultaneously. Multi-tasking is related to divided attention. Negative consequences of divided attention include increased mental effort and workload, and deterioration of performance.

- **Sustained attention:** Focusing on a specific or several elements for a prolonged period. Vigilance is related to sustained attention. Negative consequences of sustained attention are increased fatigue, loss of alertness, and deteriorated performance.

- **Attentional biases:** Biases influencing perception and attention. In stressful situations attention tends to narrow. Tunneling is basically allocating your attention to a particular channel of perception (for example, only looking or only listening), or focusing on the information for a specific task or on a specific aspect of that task. That focus is typically at the expense of the perception of other information that is not directly relevant to the attended information or task.

Cognition: The processing of information, its utilization for various purposes, and its storage for future use. Memory mechanisms play a significant role in cognition.

- **Working memory:** Temporary storage of information for immediate use in performing complex tasks. The working memory is influenced by the modality receiving the information (for example visual or auditory). It can be easily overloaded with multiple tasks, very complex tasks, time constraints, and conflicting information. Working memory is limited in its capacity and is for a short duration. Deterioration in working memory performance can degrade performance of tasks.

- **Long-term memory:** An archive of information about past events, experiences, and acquired knowledge. Information is stored better if it is rehearsed, elaborated upon, well organized, and there are relevant retrieval cues in the environment and the situation. If none of these exist, decay of information takes place (forgetting). The structure of the information in our long term memory constitutes concepts which facilitate *understanding* of the world around us. They are also the basis for mental models to be described next.

- **Mental models:** Regardless of the complexity of the situation, when people interact with other people and systems, they develop a subjective internal representation of the system and of its role in the environment. A mental model is a combination of the individual's subjective perceptions, concepts, ideas, and perceived system status. It is an organized knowledge structure that allows individuals to interact with their environment. This knowledge allows people to predict and explain the behavior of the world around them, to recognize and remember relationships among components of the environment, and to construct expectations for what is likely to occur next. Mental models are individual, but they are also shared among team members so everyone is "on the same page."

o **Problem solving:** The cognitive processes one goes through when trying to solve problems for which the solution is not obvious and not immediately available. Something that is typical to this process is the way one represents the problem. The way a person represents the problem is prone to a fixed mindset—a tendency to respond in the same way as in the past. Various problem-solving approaches are trial and error, creativity, means-end approach, or the algorithmic approach. In this respect, it is important to emphasize the differences between experts and novices: experts possess more knowledge, which is organized differently from novices. Experts spend more time analyzing problems and consequently tend to solve problems faster and with a higher success rate. Problem solving is susceptible to cognitive biases that are discussed below.

o **Decision making:** Decision making (which follows diagnosis) is choosing a solution or action that is best among other alternatives, based on the diagnosis of what is wrong with the patient. Most decisions favor the direction of increasing the utility of the choice. Decision making (and problem solving) is susceptible to biases such as the confirmation bias, the tendency to selectively assess the information that will support previous assumptions and hypotheses and ignore other information that does not support those assumptions.

o **Cognitive Biases:** Biases are cognitive errors that influence thinking, problem solving, and decision making. Anchoring is a cognitive bias in which there is a fixation on the initial information received and its initial assessment, making it unlikely to reevaluate and update the assessment with new information. Confirmation bias is the tendency to look for evidence that confirms or matches prior expectations or the current situation or decision.

Learning: The acquisition and retention of new knowledge, behavior, or skill through a process of receiving the information and practice of the behavior or skill. Learning is based on various cognitive processes described above. Some parameters of learning that are relevant to the human factors analysis of healthcare contexts and cases are the:

o duration of the training and learning;

o tools and approaches used for the learning and practice;

o decay of learning,

o provisions to re-learn and re-train (to mitigate decay), and

o amount of learning, practice, and experience required to become an expert.

Spatial Cognition: How do we know where we are and find our way to our destination? How do we learn to do that in a new environment? Way-finding is simply getting from one point to another within a given environment or geographical area. Orientation is basically the ability to know one's location within the environment and the relative location of other elements, and to continually update this knowledge (Hunt and Waller, 1999). Spatial cognition is made up of the following.

- **Perceptual processes:** perceiving and attending to features and landmarks in our environment.

- **Cognitive processes:** remembering and visualizing spatially.

- **Proprioceptive processes:** sensing relative position and strength of effort involved in movement based on information from our proprioceptive senses about our orientation and locomotion.

Situational Awareness: Put simply, "knowing what is going on." This is both a cognitive ability and cognitive processes whereby one perceives information about the situation and assesses it. There are definitions suggesting that thinking ahead is also a critical component of knowing what is going on. Situational awareness is a critical component in many situations and is linked to effective performance, on the one hand, and to adverse events when it is degraded, on the other. Situational awareness is both on the individual level and on the team level.

Motivation: Motivation consists of forces acting within or upon us to initiate and lead actions and behaviors. We also associate motivation with the intensity of our behavior and actions, whereby high motivation would be associated with more intense behavior, and vice versa. Another aspect that is associated with the level of our motivation is the persistence of the related behavior or action; that is, the relevant behavior will persist longer with higher motivation, and vice versa.

Attitudes: Attitudes are our inner tendencies toward someone or something (a physical object, an organizational process, a job or a task, a policy, etc.), and are expressed by the evaluations of that someone or something (Ajzen and Fishbein, 1977). The importance of attitudes to the human factors analysis is that various factors in the environment could influence and drive the acquisition of attitudes, which in turn, could result in various motivations and behaviors.

Anthropometrics and Action: Anthropometrics is a discipline that focuses on the measurements of human limbs, positions, and actions, and the relations of those measurements to their environment and tools. There are many fields of study that focus on anthropometrics, including ergonomics, biomechanics, and occupational health. There are various standards for anthropometrics for various purposes (e.g., ISO 7259). That data can help us assess aspects in the physical environment with respect to space and layout. We should assess anthropometrics in terms of the actions one takes when performing tasks.

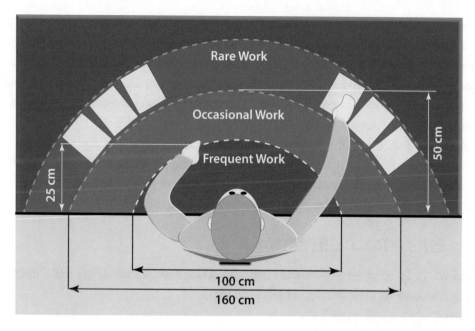

Figure 7.6: Anthropometric data showing workspace. Adapted from the guideline of the U.S. OSHA guidelines.

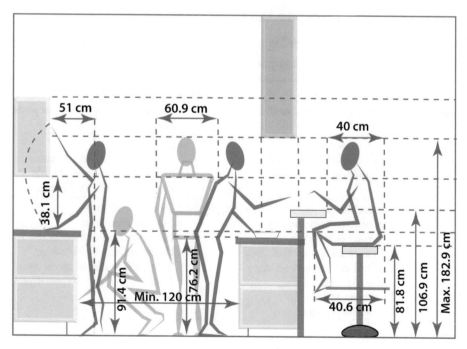

Figure 7.7: Anthropometric data showing hands reach envelopes for various purposes. Adapted from Tilley (1993).

Anthropometric data helps us assess the ability to reach somewhere (for example, a nurse taking a medication from the top shelf in a cabinet in the medication room, or bending down to take medication from a bottom drawer that is close to the floor); or the amount of space one has in order to perform tasks, and interact with others. Other anthropometric aspects to consider are one's physical abilities, such as gross and fine motor skills, physical strength, and condition of the respiratory and cardio-vascular system.

Executive Functions: There are processes of allocating mental and attentional resources to manage perception, and cognition, and action. These mental resources are limited, and this limitation is typically the reason for the difficulties we sometimes face when being overloaded. This limitation will become a major consideration in the assessment of fit.

7.2   ASSESSMENT CRITERIA

How can we apply what we understand about the above human capabilities and limitations to assess the fit? Consider these simple, yet powerful steps.

1. Use the mapping we did.

2. Consider the human capabilities and limitations relevant to the task and environment.

3. Assess fit using the detailed checklist for each of the factors.

Table 7.1 is a general table outlining the capabilities relevant to the major mapping categories along with some examples of finding poor fit that reflects problems.

Table 7.1: Human capabilities relevant to the major mapping categories			
	Mapping	**Human factors to consider**	**Fit Assessment (looking for problems) Examples**
Goals and Tasks	Goals	Understanding Adequacy of mental models Attitudes and motivation	Goals not clear Lack of goals in mental models Negative attitude toward goals Poor motivation to achieve goals
	Tasks	Task difficulty (cognitive—memory, physical, and required accuracy) Number of tasks Time pressure to perform tasks Attention and multi-tasking Mental model Learning and practice Executive function	Task too hard/easy Too many/few tasks Too much/little time pressure Too much/little multi-tasking Inappropriate mental model about how to do the task

	Mapping	Human factors to consider	Fit Assessment (looking for problems) Examples
Physical Environment	Physical space	Visual perception (search, visual field, and visual acuity) Orientation and way-finding Executive functions Anthropometry (for example, reach) Minimal personal space Action: walking distance	Things are not visible, out of the visual field, or in the periphery Space too confusing Things are out of immediate reach and require movement (for example, walking, stretching, or bending) Below minimal required space per person Too far to get to required places or items
	Layout of sub-spaces and general locations	Visual perception (that is, search, visual field, and visual acuity) Attention Spatial cognition Mental models and concepts Action: Reach envelope (that is, within arms reach) Action: walking	Required places not within field-of-view (FOV) or line-of-sight Factors not supporting visual search: Not organized or coded adequately Desired elements out of FOV Poor legibility Distractors to attention Lack of support for spatial orientation Distances too far from required destinations (for example, reaching, bending, or walking)
	Artifact and device locations and access	Detect-Recognize-Identify-Reach-Take Visual (search, visual field, visual acuity) Attention Spatial cognition Mental models and concepts Executive functions Action: Reach envelope Action: Walking	Required artifacts and devices not within field-of-view (FOV) or line-of-sight Too many distracters Spatial layout confusing Layout not as expected Taking or placing artifacts requires additional actions (extra reach, bending, walking)
	Ambient conditions	Light requirements Noise limits Temperature limits	Not enough light, reflections Too much noise Temperature too cold or too hot Too humid or too dry
	User Interface (UI) of devices and tools [Also known as Human Machine Interface (HMI) and Human Computer Interface (HCI)]	Mental model Training and learning Usability and good usability heuristics (see Table 7.3 below)	Poor user interface concept Lack of training Poor feedback in user interface Not enough flexibility

	Mapping	Human factors to consider	Fit Assessment (looking for problems) Examples
Human Environment	Teams and groups	Team training Team skills Team composition and structure Leadership	Lack or insufficient team training Team members lack team skills Team structure does not support team task and workflow Poor or lack of leadership
	Organizations and or-ganizational culture	Organization structure Employer-employee relationships Management and leadership Workload management Information flow and organizational learn-ing Service culture Safety culture	Organizational structure does not support goals and tasks Little employer or employee commitment Workload managed poorly Poor information flow within the organi-zation Organization is not patient-oriented No safety campaigns or training
	Rules, reg-ulations, and poli-cies	Organization policies Clinical procedures and good practices	Lack of rules, regulations, and policies Unclear rules, regulations, and policies Poorly written, implemented, understood, and/or maintained policies and procedures

The following sections will guide you through the detailed process of assessing the fit.

7.3   PEOPLE AND TASKS

Our initial concern is the fit between people and their tasks. We do this assessment from two perspectives, which are two sides of the same coin: characteristics of the people and their fit to the tasks, and characteristics of the tasks and their fit to people.

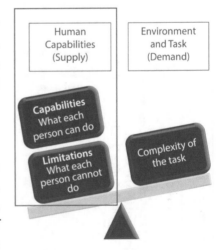

7.3.1   THE PEOPLE PERSPECTIVE

We start the assessment with the people perspective. In other words, we focus on the unit of analysis you address, for example, an individual or a team, their characteristics, and how those fit the task. This is an overall assessment of several aspects: experience and qualifications, motivation and attitudes, decision-making capabilities, risk awareness and risk taking, and handling of stress.

Use the detailed checklist and consider the following questions for the appropriate unit(s) of analysis (for example, an individual person, people in a group or team, or an organization).

1. Do the people have the relevant training, certification, and experience to perform the tasks?

2. Is the person or are the people in the team motivated to do their job?

3. Do they have positive attitudes toward safety and quality?

4. Are the people aware of possible risks and hazards in their work place?

5. Do the people tend to avoid taking unnecessary risks while performing their tasks?

6. Do the people have a perception that they are in control of what is happening?

7. How do the people handle stress?

8. Are people nervous or relaxed?

9. Can people divide their attention between other people and their tasks?

10. How do people make the decisions required to accomplish their tasks? Do they make the decisions effectively?

11. Do people seem to exhibit any cognitive biases?

The following answers assess the fit of the relevant people to tasks in the sepsis management scenario.

Checklist Item	Data	Fit Score
1	The nurses may not have a good knowledge and appreciation of the management of septic patients, particularly in recognizing a sepsis patient.	4
2	By and large, motivation is high.	7
3	In general, the nurses are committed to providing high quality care.	6
4	The nurses are aware of some issues in their work environment. However, they may get used to some hazards over time.	4
5	The nurses usually don't take unnecessary risks knowingly. However, they may unknowingly get used to unsafe practices attempting to improve efficiency, or through other mechanisms.	5
6	For the most part, the nurses feel they are in control. A busy environment may change that.	5

7	In general, the nurses handle stress well.	6
8	In general, the ED nurses are nervous.	6
9	Usually the nurses can divide their attention.	6
10	The nurses make decisions mostly intuitively, or in a rule-based fashion, and usually effectively.	5
11	In the context of sepsis, cognitive biases may be common. Confirmation and anchoring biases may contribute to misdiagnosis of sepsis, and thus to failure to manage it properly.	3

7.3.2   THE TASK PERSPECTIVE

We continue with the assessment of the tasks. In this assessment, we examine various characteristics of the tasks and the extent of their fit to human capabilities and limitations. The main task characteristics and demands that we assess are as follows:

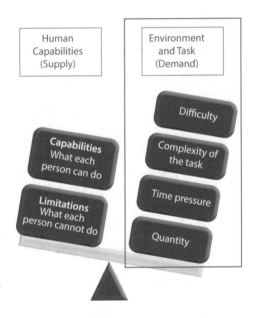

- **Quantity:** The number of tasks people are expected to perform.

- **Time pressure:** The time constraints imposed on task performance.

- **Complexity:** The number of elements and alternatives in the task.

- **Difficulty:** The scope of mental and physical resources that people have to invest in order to perform the task.

Use the detailed checklist to assess task fit for the appropriate unit(s) of analysis (for example, individual person, people in a group or team, organization):

1. Are there too many tasks for the individual or team?

2. Are they required to perform too many tasks simultaneously?

3. Is the time available to perform tasks too short?

4. Are the tasks too complex? Too difficult? Too easy?

5. Is there too much information?

6. Is the information too complex to receive, retain, or use while performing tasks?

7. Can people divide their attention between several tasks?

8. Can the user track or monitor information for a prolonged period?

9. Can people make decisions given the time they have, the number of tasks they need to deal with, and/or the amount of information they need?

10. Is the required physical effort appropriate to the people?

11. Do people exert physical effort for prolonged periods?

12. Do people exert physical effort that challenges the respiratory or cardio-vascular system?

The following answers assess the fit of the tasks to the relevant unit(s) of analysis (primarily nurses) in the sepsis management scenario.

Checklist Item	Data	Fit Score
1	The number of tasks is manageable.	7
2	Especially at times of peak activity, nurses may need to address too many tasks concurrently. Usually, however, this is not a problem.	6
3	There is sufficient time to perform the task, once the need is recognized.	7
4	The tasks are quite straightforward and easy to execute.	7
5	The necessary information is quite basic and easy to obtain.	7
6	For the most part, dividing attention to the tasks at hand is quite easy.	6
7	Focusing attention on the relevant information could be challenging at times.	4
8	Technically, it is easy to track and monitor information over time. The problem is more on the side of remembering to do so.	6
9	Technically, the time available is sufficient for making adequate decisions.	6
10	N/A	
11	Physical effort is appropriate.	7
12	Physical effort does not challenge the respiratory or cardio-vascular system.	7

7.4   PHYSICAL ENVIRONMENT

Now we add another layer to the assessment by looking at the fit between characteristics of the physical environment and the people and their tasks.

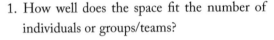

7.4.1   SPACE AND LAYOUT

In assessing the fit between the people and tasks on the one hand, and the physical environment on the other, we begin with the space and layout information you collected while mapping the environment in the first step of the project (see Section 6.3.1).

Use the detailed checklist to assess the fit of the space and layout with the tasks and people (in terms of the appropriate unit(s) of analysis, for example, individual person, people in a group or team, organization).

1. How well does the space fit the number of individuals or groups/teams?

2. Is there sufficient space for the individual or group/team to perform tasks and achieve goals?

3. Are the sub-spaces configured in a way that supports the workflow to be effective, efficient, and safe?

4. Are the access paths and routes convenient?

5. Are the entry and exit points located where they support the workflow?

6. Are the locations of doors and windows or any other fixed elements appropriate?

The following answers assess the fit of the space and layout to the people (the relevant unit(s) of analysis—primarily nurses) and the tasks in the sepsis management scenario.

Checklist Item	Data	Fit Score
1	With recently increasing crowding in the ED, it is not uncommon that space is limited.	5
2	For the most part space is sufficient to perform tasks.	6
3	Space configuration can pose challenges to the workflow. It may be difficult to see a patient; patients need to be transferred from the triage area to the main ED space to be treated.	4
4	Quite often access paths are crowded.	4
5	Entry and exit point locations are less relevant to the sepsis workflow.	7

For this part of the assessment, in addition to the data about the location of elements in the physical environment, we use the link or spaghetti diagram that you have sketched as part of the space and layout mapping (refer back to Figure 6.6). Examine the links or the spaghetti in terms of their characteristics, such as:

- frequency of the association between the elements;

- operational sequence;

- importance of the associations between the elements; and

- length and complexity (for example, do they depict a twisted route, cross many other links, support a line of sight, and support a line of communication?).

7.4.2   LOCATION OF ARTIFACTS AND DEVICES

Use the detailed checklist to assess fit of the locations of artifacts and devices with people (in terms of the appropriate unit(s) of analysis, for example, individual person, people in a group or team, organization) and their tasks.

1. Are the artifacts one uses for performing tasks located appropriately for purposes of reaching, viewing, and/or hearing?

2. Are the artifacts located appropriately for several people to perform their tasks?

3. Does the layout support people in focusing on information or elements they require for their tasks?

4. Can people notice and receive from their location relevant information even though they are not focusing on it?

5. Can people divide their attention between several sources of information?

6. Is information displayed visually in a way that supports reading and identification of details?

7. Are all required information and artifacts within the field of view of the people who need it? If not, does it require much eye and head movement to perceive it?

The following answers assess the fit of the space and layout to the people (the relevant unit(s) of analysis—primarily nurses) and the tasks in the sepsis management scenario. The number beside each answer is the fit score.

Checklist Item	Data	Fit Score
1	Vital sign measurement devices, as well as the sepsis workbook, are located at the triage booths.	6
2	Sepsis workbook is not at point of care; However, location of other artifacts less relevant to this scenario, as it is uncommon for the triage nurses to encounter multiple septic patients concurrently.	6
3	It is very easy to lose track of a patient, and his/her information.	2
4	Lack of focused attention in the visual search could result in failure to identify important information.	2
5	It is not very difficult to divide attention between the sources of information. The issue is focusing attention on the right information.	5
6	Information per se is displayed legibly. However, the nurse needs to initiate seeking information.	5
7	Nurses often need to get up from their chair and move around the space to get necessary information and artifacts.	2

7.4.3   AMBIENT CONDITIONS

Use the detailed checklist to assess fit of the ambient conditions with people (in terms of the appropriate unit(s) of analysis, for example, individual person, people in a group or team, and/or organization) and their tasks.

1. Is there sufficient light to perform the task? Too much? Too little? Are there too many reflections? Are the reflections disturbing?

2. Are the lighting conditions such that they do not require adaptation to low-level lighting?

3. Is the level of the noise reasonable?

4. Does the noise create distractions and interruptions?

5. Can people hear all the auditory information they require?

6. Can they distinguish between sounds (for example, alarms) and speech?

7. Is speech intelligible?

8. Are they required to localize the source of sounds or speech, and can they do it?

9. Is the temperature in the work environment appropriate? Not too cold? Not too hot?

10. Is the air quality in the work environment appropriate?

11. Are there vibrations in the work environment? If yes, does the frequency or intensity of vibrations create disturbance?

12. Do the odors create a problem (discomfort, breathing difficulty, tearing, and any other responses)?

The following answers assess the fit of the ambient conditions to the people (the relevant unit(s) of analysis—primarily nurses) and the tasks in the sepsis management scenario. The number beside each answer is the fit score.

Checklist Item	Data	Fit Score
1	Light is adequate; there are some reflections from glass windows and partitions.	6
2	Lighting conditions not an issue.	7
3	The triage area can be noisy; distractions and interruptions are common.	4
4	For the most part, noise distractions and interruptions are not a problem.	6
5	While distinguishing between different sounds may be challenging in the triage area, this is less of an issue for the sepsis management task.	6
6	N/A	
7	N/A	
8	N/A	
9	The area is air conditioned and comfortable.	7
10	N/A	
11	N/A	
12	Intense odors are uncommon.	6

7.5   DEVICE USABILITY

The appropriateness or inappropriateness of the user interface (UI) design and implementation is often assessed and tested in terms of its usability. The usability of artifacts and devices could be critical to performance and safety. In assessing the fit with devices, we perform a usability evaluation. In a usability evaluation, we evaluate to what extent the device enables users to achieve their goals in terms of (ISO 9241-11):

- effectiveness: success in completing tasks and achieving goals in a safe manner;

- efficiency: resources expended in order to complete tasks and achieve goals; and

• user satisfaction: the emotional and experiential aspects of the interaction.

Note that there are FDA guidelines and recognized standards for the design and usability of medical devices. Table 7.2 shows the standards that the FDA recommends.

Table 7.2: FDA-recommended standards

Standard	Title	Main Purpose
AAMI/ANSI HE75:2009	Human Factors Engineering—Design of Medical Devices	Comprehensive reference that incudes general principles, management of use error risk, design elements, integrated solutions
ANSI/AAMI/IEC 62366-1:2015	Medcical Devices—Part 1L Application of usability engineering to medical devices	HFE/UE process applied to all applying HF/usability to medical device design, with consideration of risk management
ANSI/AAMI/ISO 1497 1:2007/(R)2010	Medical Devices—Application of risk management to medical devices	Risk management process for medical devices
IEC 60601-1-6:2010	Medical Electrical Equipment—Part 1-6: Feneral requirements for basic safety and essential performance—Collateral standard: Usability	Provides a bridge between IEC 60601-1 and ANSI/AAMI/IEC 62366
IEC 6061-1-8 Edition 2.1 2012-11	Medical Electrical Equipment—Part 1-8: General requirements for basic safety and essential performance—Collateral Standard: General requirements, tests, and guidance for alarm systems in medical electrical equipment and medical electrical systems	Design standard for alarm systems in medical electrical equipment and systems
IEC 60601-1-11:2010	Medical Electrical Equipment—Part 1-11: General requirements for basic safety and essential performance—Collateral Standard: Requirements for medical electrical equipment and medical electrical systems used in home healthcare environment	Reguirements for medical electrical equipment used in non-clinical environments, including issues involving medical device use by lay users

Taken from: http://www.fda.gov/MedicalDevices/DeviceRegulationandGuidance/HumanFactors/ucm119190.htm.

The following is a simple checklist covering the key aspects in the user interface of a device, following the ten usability heuristics suggested by Nielsen (1993).

Table 7.3: The ten usability heuristics suggested by Nielsen (1993) along win and agreement-disagreement rating scale

Statement	Strongly Disagree			Undecided			Strongly Agree
	(1)	(2)	(3)	(4)	(5)	(6)	(7)
1. The work flow has a clear beginning and end.							
2. All functions are grouped to support the work flow.							
3. The device keeps the user informed about what is going on.							
4. The information appears in natural and logical order.							
5. The device allows the user to have full control.							
6. The device prevents errors from occurring.							
7. The device provides clear ways to undo actions and recover from errors.							
8. The device "speaks" the users' language.							
9. The device looks aesthetically pleasing.							
10. The devicet has a simple visual design.							

Table 7.4 evaluates the device usability of the electronic health records (EHR) for the sepsis management scenario.

Table 7.4: Device usability of the Electronic Health Records (EHR) for the sepsis management scenario

Statement	Strongly Disagree			Undecided			Strongly Agree
	(1)	(2)	(3)	(4)	(5)	(6)	(7)
The work flow has a clear beginning and end. The flow of tests is not clear and requires searching for them; they are not pushed to the user. Delays in flow of information from the room into the Emergency Department Information System.				✓			
All functions are grouped to support the work flow. User has to interact with several different systems. The systems are very different; emergency has another system for X-Ray results.				✓			
It keeps the user informed about what goes on.				✓			
The information appears in natural and logical order.				✓			
It allows the user to have full control.				✓			
It prevents errors from occurring. There could be omission errors.				✓			
It provides clear ways to undo actions and recover from errors.				✓			
It "speaks" the user language. Emergency Department Information System (EDIS) gives the blood test results without trends, only numbers; trend is very important, and for that one needs to log into another system.				✓			
It looks aesthetically pleasing.				✓			
It has a simple visual design. Color coding of monitor may not be clear to everyone.				✓			

7.6   HUMAN ENVIRONMENT

Does the human environment support people and facilitate performing tasks and achieving goals?

Answering this question is the main objective of assessing the fit of various factors in the human environment. Following the mapping of the human environment, we perform the assessment on three key aspects.

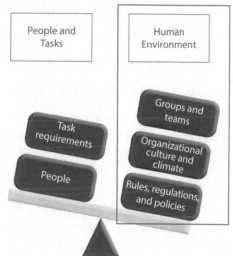

1. Groups and teams.

2. Organizational culture and climate.

3. Rules, regulations, and policies.

7.6.1   GROUPS AND TEAMS

Remember, we address groups and teams as part of the human environment, when our unit of analysis is the individual. Use the detailed checklist to assess fit with groups and teams, and assess teamwork.

1. Does the presence or absence of any of the people in the workspace support or disrupt the workflow and tasks?

2. Can people divide their attention between other people and their tasks?

3. Does the structure of the team support teamwork and team tasks and goals?

4. Is there a leader? Is the leadership effective?

5. Do all team members share recognition/familiarity with the professional expertise or authority of each member in the team?

6. Do team members communicate effectively?

7. Do they share mental models?

8. Is there sharing of awareness and understanding of the situation (situational awareness)?

9. Does communication become disruptive (does it become an interruption or distraction)?

10. Is there effective coordination and collaboration?

11. Is there a clear definition of roles?

12. Is there a clear and fair allocation of workload?

13. Do people have good structured knowledge of who does what, when, and how (mental models)?

14. Do team members share similar expectations and understanding of everyone's role and what is supposed to be done, when, and how?

15. Are people capable of being aware of what is going on? Do they share this knowledge?

16. Do team members effectively share awareness and understanding of the situation?

The following answers assess the fit of the space and layout to the people (the relevant unit(s) of analysis—primarily nurses) and the tasks in the sepsis management scenario.

Checklist Item	Data	Fit Score
1	Presence of people is usually not an issue. Occasionally, in a very busy environment, workflow may slow down.	6
2	During usual workload nurses can divide attention well between relevant people and tasks.	6
3	The team at the triage area is quite small, consisting of 1–2 nurses and 1–2 clerks.	6
4	It is unclear whether a leader has been assigned in the triage area.	4
5	Nurses attend only to nursing tasks; clerks attend only to their own tasks.	1
6	It is hard to tell whether communication is effective. There are no clear expectations for communication, and decisions are left in the hands of members.	4
7	It is not clear if they have shared mental models.	4
8	It is not clear if they have shared situational awareness.	4
9	The challenge is more in failing to communicate.	7
10	Once an issue is communicated, there is pretty good coordination and collaboration among members.	6
11	There is a clear definition of roles.	7
12	The allocation of workload can be a hit or miss. When the environment is busy, typically everyone would work harder.	6
13	People have good structured knowledge of who does what, when, and how.	7
14	Team members share similar expectations and understanding of everyone's role and what is supposed to be done when and how.	7
15	Team members are aware of what is going on and sometimes communicate it.	6
16	It is easy to lose situational awareness and not share it, especially in a busy environment.	4

7.6.2   ORGANIZATIONS, CLIMATES, AND CULTURES

Use the detailed checklist to assess the fit with organizational culture and climate.

1. Is the division and definition of roles within the organization clear?

2. Does the organizational structure support the achievement of its goals?

3. Is the staffing sufficient or appropriate?

4. Is the organization committed to its members in a way that promotes positive attitudes and behaviors?

5. Are members of the organization involved and engaged?

6. Does the organization have a positive culture with respect to safety and quality?

7. Does the organization have an accountability culture that is positive, supportive, and constructive?

8. Does learning take place in the organization in a way that professional development and changes are supported?

9. Is there an effective transfer of information within the organization?

10. Is available training and certification sufficient or appropriate for people to perform their tasks?

The following answers assess the fit of the organization, climate, and culture to the people (the relevant unit(s) of analysis—primarily nurses) and the tasks in the sepsis management scenario.

Checklist Item	Data	Fit Score
1	Division of roles within the organization is clear.	7
2	The organization structure supports achieving the goals.	7
3	Staffing is appropriate.	7
4	Employee surveys have revealed some degree of dissatisfaction among staff regarding support from leadership.	4
5	Members of the organization are involved and engaged for the most part.	6
6	Safety culture surveys have repeatedly indicated suboptimal perceptions of staff regarding safety and quality.	4
7	The term "accountability" often has a negative connotation, implying "you better do your job!"	3

8	Learning toward improved practice is uncommon.	2
9	Communication is based on intranet/email, which fail to reach many.	2
10	Generally, team members are adept in resuscitation. The missing training aspect is in recognizing a septic patient.	6

7.6.3   RULES, REGULATIONS, AND POLICIES

Use the detailed checklist to assess the fit with rules, regulations, and policies.

1. Are there sufficient or appropriate policies?

2. Does the organization rely too much on rules and procedures?

3. Are the policies designed and communicated in a way that supports comprehension, even in stressful situations?

4. Are the policies designed and communicated in a way that supports adherence?

5. Are the policies current and refreshed?

6. Are policies and relevant resources available and accessible?

7. Are there too many workarounds? If yes, what is the reason?

8. Is the allocation and distribution of workload appropriate? Consider work hours, shifts, rotations, rest periods, night work.

The following answers assess the fit of the organization, climate, and culture to the people (the relevant unit(s) of analysis—primarily nurses) and the tasks in the sepsis management scenario.

Checklist Item	Data	Fit Score
1	There are sufficient and appropriate policies.	7
2	There are very limited activities to ensure that policies, rules, and procedures are actually followed.	1
3	Policies are not designed and communicated in a way that supports adherence.	1
4	Policies are current for the most part.	6
5	Policy resources are available; however, staff usually doesn't seek to find and read them.	4
6	It is unknown if there are too many workarounds. In the case of sepsis, workarounds are likely not a substantial issue.	5
7	For the most part workload is appropriate.	6

CHAPTER 8

Interim Findings: Problems of Fit

Now that we have assessed the extent of fit between task demands and the environment, on the one hand, and human capabilities, on the other, we are ready to put together our second summary: the fit problems and their severity. Note that while we focus on problems, we should not forget about factors that have a good fit that we want to preserve or facilitate.

8.1   LIST OF THE FIT PROBLEMS BY FACTORS

We propose summarizing the fit problems by the factors in the foundation tier of the human factors framework. Any mapped factor with poor fit, in the red zone on the scale below, can be considered as a fit problem. Any mapped factor with medium fit, in the yellow zone on the scale below, should give us some concern. Any mapped factor with good fit, in the green zone on the scale below, should be further facilitated.

1	2	3	4	5	6	7

Poor fit   Good fit

Briefly state the fit issue with each of the factors. For example, with a unit of analysis of a team of three people and the physical space factor, one may state that the available space is insufficient for the three-person team to perform their tasks, with a fit score of 3 out of 7 in the fit scale. Note that the summary is for each relevant unit of analysis, which could be an individual, a team, or even an organizational unit. Table 8.1 can be used for summarizing the fit issues and Table 8.2 is an example of the summary of fit assessment for the sepsis management scenario.

Table 8.1: Strategy for estimating problem severity

Unit of Analysis	Context		Fit with capabilities and limitations	Fit score
	Factors	Sub-factors		
	Goals and Tasks			
	Physical Environment	Physical space		
		Layout of sub-spaces and general locations		
		Artifact and device locations and access		
		Ambient conditions		
		HMI of devices and tools		
	Human Environment	Teams and groups		
		Organizations and organizational culture		
		Rules, regulations, and policies		

Table 8.2: Sample summary of fit assessment for the sepsis management scenario

Unit of Analysis	Context		Fit with capabilities and limitations	Fit score
	Factors	**Sub-factors**		
Triage area, including a unit clerk and a triage nurse	Goals and Tasks		The task of identifying and initiating the workflow for septic patients has a good fit with the capabilities of involved team members. However, it is quite likely that nurses often lack the expertise to recognize a septic patient, especially in early stages of illness. Focusing on the right information is challenging. That said, providers are highly motivated to do a good job. Cognitive biases may occur.	4
	Physical Environment	Physical space	The area is crowded.	4
		Layout of sub-spaces and general locations	The physical space and crowdedness pose challenges to providers' ability to observe and monitor a septic patient.	3
		Artifact and device locations and access	The sepsis workbook (protocol) is unavailable at the point of care; nurses often need to exert extra effort to access relevant information.	3
		Ambient conditions	Although crowding and noise can be a nuisance at the triage area, the ambient conditions are comfortable for the most part.	5
		HMI of devices and tools	The logical grouping and consistency of information relevant to sepsis in the EHR is inappropriate.	4
	Human Environment	Teams and groups	There is no clearly defined leader. It is not clear if communication is always effective, if there is a shared understanding at all times, and if situational awareness is degraded.	3
		Organizations and organizational culture	There is some dissatisfaction with leadership. The safety climate is insufficient. An "accountability" culture is perceived negatively. There is insufficient learning to improve. Within an organization communication does not reach everyone.	3
		Rules, regulations and policies	Mechanisms to ensure adherence to policies and procedures are insufficient. Policies are not implemented appropriately. Staff do not always seek the relevant policies.	4

8.2   PROBLEM SEVERITY

The next step is to assess the severity of the fit problems. You may wonder: Why assess severity if we already have the fit score? Let us first address "severity" and then get back to this question.

The criteria for assessing the severity are

1. frequency of the problem or likelihood of the problem to occur and

2. impact of the problem on patient safety, care quality, operations efficiency, patient experience, and satisfaction.

When we assess severity with parameters such as frequency and impact, we know that sometimes there could be problems that may occur rarely or have very little impact, and we refer to those as having low severity. In a similar way, there could be problems of fit, but those problems could either be highly improbable or have a negligible impact. For this reason we aim to not only assess the fit, but also assess the severity of fit problems. Later on, we will focus on the more severe problems to recognize emergent factors, and go on to conclude about the most influencing factors on performance and outcomes.

Table 8.3 provides a strategy for assessing the severity of the fit problems. Use the value for each problem area to complete Table 8.4, summarizing the problems and describing them along with their severity as a function of the key factors.

1. Assess the frequency in terms of how high or low the probability of occurrence is. For highly probable, use 1. For improbable, use 7. Choose the number that best reflects your assessment of the probability of occurrence based on the frequency of occurrence. The more frequently the problem occurs, the higher the probability of occurrence.

2. Assess the impact of the problem in terms of how negligible or irreversible it is. Impact is connected with the criticality factor described in Section 6.2.2 which discusses implications of the task analysis to severity ratings.

3. Identify the cell where the probability of occurrence and impact meet. Low severity problems are color-coded green. Medium severity problems are color-coded yellow. And high severity problems are color-coded red. Both the medium and high severity problems should concern us.

Table 8.3: Strategy for estimating problem severity

			Impact						
			Negligible						Irreversible
			1	2	3	4	5	6	7
Probability of Occurrence	Highly Probable	1	Medium	High	High	High	High	High	High
		2	Medium	Medium	High	High	High	High	High
		3	Medium	Medium	Medium	High	High	High	High
		4	Low	Medium	Medium	Medium	High	High	High
		5	Low	Low	Medium	Medium	Medium	High	High
	Improbable	6	Low	Low	Low	Medium	Medium	Medium	High
		7	Low	Low	Low	Low	Medium	Medium	Medium

Table 8.4: Problem summary table

Key Factors	Problem area	Problems	Severity
Goals and Tasks	Goals		
	Tasks		
Physical Environment	Physical space		
	Layout of sub-spaces and general locations		
	Artifact and device locations and access		
	Ambient conditions		
	Usability of devices and tools		
Human Environment	Teams and groups		
	Organizations and organizational culture		
	Rules, regulations, and policies		

Table 8.5 shows an example of the summary of severity of problems in the sepsis management scenario.

	Problem area	Problems	Severity
Table 8.5: Summary of severity of problems in the sepsis management scenario			
Goals and Tasks	Goals		
	Tasks	Variable level of skill among nurses in recognizing that a patient is septic: Fairly common; when present likely to be a problem with major impact on outcomes	High
Physical Environment	Physical space	Crowdedness	Medium
	Layout of sub-spaces and general locations	Challenges to providers' ability to observe and monitor a septic patient	Medium
	Artifact and device locations and access	Sepsis workbook not present: Uncommon; when missing, can be accessed somewhere in the ED; good clinical skills can negate the need for it.	Low
		Lack of readily available information, and alerts regarding, abnormal physical and lab-based findings: Common, substantial impact on outcomes.	High
	Ambient conditions	Noise	Low
	Usability of devices and tools	Inconsistent and hard to reach information	Medium
Human Environment	Teams and groups	Once sepsis is recognized, team members communicate well regarding sick patients. However, failing to close communication loops may have a negative impact, and with substantial consequences	High
	Organizations and organizational culture	Poor capability to promote learning in the organization can have major consequences	High
	Rules, regulations, and policies	Problem with adopting and adhering to policies and regulations	Low

CHAPTER 9

Recognize Emergent Factors

9.1 DEVELOPMENTS THAT MAY INFLUENCE THE PERFORMANCE AND OUTCOMES

Having completed the fit assessment, we already know a lot about the situation and the factors that have likely influenced performance and outcomes, or that are likely to be influential if we are dealing with a proactive scenario. However, additional factors that can influence performance and outcomes may emerge because of the extent of fit between the demands of the task and environmental factors, on the one hand, and human capabilities, on the other hand. Moreover, some additional environmental factors can also emerge as a result of the inter-play between some of the environmental and contextual factors. Before we conclude that performance and certain outcomes are due to the problems we have uncovered based on our mapping and assessment, we should recognize the additional emergent factors that may further influence performance and outcomes.

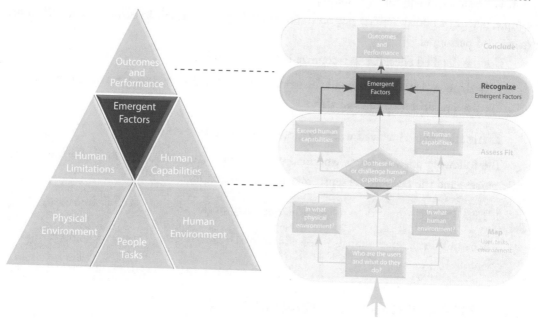

Figure 9.1: Recognize the emergent factors.

There can be two types of emergent factors:

1. environmental (physical and human) and

2. human (psychological, physical).

9.2   EMERGENT ENVIRONMENTAL FACTORS

New factors in the physical and/or human environment can emerge due to conflicts or tensions among the factors that were in the situation, process, or site to start with.

Follow the checklist of questions to guide you in recognizing emergent factors in the various categories. You should also assign a severity score to each of the emergent factors you recognize. The criteria for assigning a severity score should be similar to the criteria we used earlier for the problems (see Table 8.1).

9.2.1   WORKLOAD

Workload is the amount of work one is assigned and/or actually performs. It could be an emergent factor in the human environment due to problems in any of the factor categories in that environment. For example, inappropriate staffing could cause more work for some of the staff. Another example, placing a poorly trained person to do a certain job can result in extra work for the more experienced personnel.

Use the detailed checklist to recognize increased workload.

1. Is there an increased demand for work that is beyond the normal, to any of the involved healthcare professionals?

2. Are the people involved working more than normal?

The following answers assess workload of the people (the relevant unit(s) of analysis—primarily nurses) and the tasks in the sepsis management scenario.

1. Sepsis management does not typically increase workload significantly.

2. The healthcare professionals involved in sepsis management do not work significantly more than normal.

9.2.2   DISTRACTIONS AND INTERRUPTIONS

A distraction is any event in the work environment that can distract one's concentration and focus on their primary task, but it does not require them to shift their attention from the primary task. Distractions can slow down and degrade performance of the primary task.

An interruption is any event that takes an individual or a team away from their primary task and requires them to temporarily attend to and deal with another task. A common example in healthcare is physicians getting phone calls or being paged urgently. They need to attend to the interruption and then get back to their primary task. The costs of interruptions are possible loss of awareness to the main task and anything in the situation that is related to the task; delay in resuming the primary task, and the possibility or omitting required actions.

Follow the checklist to help you recognize the emergent distractions and interruptions.

1. Does the work environment introduce distractions or interruptions?

2. Are the distractions or interruptions frequent?

3. Do people seem to be affected by the interruptions or distractions?

The following answers assess distractions and interruptions of the people (the relevant unit(s) of analysis—primarily nurses) and the tasks in the sepsis management scenario.

1. The work environment does introduce distractions and interruptions.

2. The distractions and interruptions could be high at times.

3. Using the sepsis protocol can degrade the impact of the distraction and interruptions. If not, there could be adverse impact of the distractions and interruptions.

9.3   EMERGENT HUMAN FACTORS

9.3.1   MENTAL AND PHYSICAL WORKLOAD

Workload in this context is the amount of resources—physical or mental—one has to invest relative to the requirements and demands of performing the task, given all other contributing factors (environmental factors as well as human capabilities and limitations). It should be noted that we often talk in terms of the perceived or experienced workload in this context. If the investment of resources fits the demands, then the workload is manageable. However, if demands exceed available resources, then a high workload is perceived and experienced. Workload can be physical and/or mental. High workload could result in performance deterioration, and also in physical and emotional harm; it could lead to harmful stress levels.

Mental Workload

We suggest adopting the definition by Hart and Staveland (1988): "the perceived relationship between the amount of mental processing capability or resources and the amount required by the task."

Use the detailed checklist to recognize increased mental workload.

1. Do the perceived cognitive demands of the task or the environment exceed one's cognitive capabilities?

2. Which cognitive capabilities are overloaded (exceeded)?

The following answers assess the mental workload of the people (the relevant unit(s) of analysis—primarily nurses) in the sepsis management scenario.

1. Cognitive capabilities are not overloaded, particularly when using the sepsis protocol. Not using the protocol may add substantially to providers' cognitive load in having to remember the different management elements, and following up on the different indicators of a patient's response to treatment. However, sometimes the lack of relevant skills in recognizing sepsis may increase mental workload, particularly in the subsequent procedure once it is recognized. In addition, ineffective communication may also increase mental workload.

2. Some challenges to cognitive capabilities can rise due to the unclear and inconsistent presentation of relevant information in the EHR.

Physical Workload

Physical workload is a measureable amount of physical effort expended when you perform a given task (for example, lifting or pushing, or sitting in the same place for a very long time) and is affected by a range of factors.

Use the detailed checklist to recognize physical workload.

1. Do the physical demands of the task or the environment exceed one's physical capabilities?

2. What specific physical actions are overloaded?

The following answers assess the physical workload of the people (the relevant unit(s) of analysis—primarily nurses) in the sepsis management scenario.

1. Typically, physical demands of managing sepsis do not exceed physical capabilities.

2. Due to the layout, some physical demands may increase when trying to follow and monitor the patient.

9.3.2 DISCOMFORT

It is important to recognize additional emergent factors as a function of the physical aspects of the environment.

Use the detailed checklist to recognize emergent discomfort.

1. Can the available space result in physical discomfort, pain, or harm?

2. Can any other factor result in physical discomfort, pain, or harm?

The following answers assess the discomfort of the people (the relevant unit(s) of analysis - primarily nurses) in the sepsis management scenario.

1. No discomfort results from being engaged in sepsis management.

2. The difficulty in accessing relevant information or the sepsis workbook can result in some frustration

9.3.3 FATIGUE AND LOSS OF VIGILANCE

High workload, particularly prolonged high workload, can result in the build-up of fatigue. Fatigue is the physical or mental state when there is insufficient capacity or energy to perform the required activities (physical or mental). Fatigue is also the result of prolonged sustained activity (working long hours), sleep disruption or deprivation, and sleep/rest at inappropriate times. Increased fatigue can degrade cognitive performance, among others problems.

Lack of activity, which can be associated with little or no workload, can result in loss of vigilance. When you recognize possible loss of vigilance as an emergent factor, make sure it is relevant to your case. The need for vigilance, or sustained attention, is typical to monitoring tasks.

Use the detailed checklist to recognize emergent fatigue and loss of vigilance.

1. Is one required to sustain a high workload or stressful situation for a prolonged period?

2. Are there symptoms of fatigue?

3. Is one required to sustain a very low workload for a prolonged period?

The following answers illustrate that, sometimes, the questions may not be applicable, as in the case for assessing the fatigue and loss of vigilance of the people (the relevant unit(s) of analysis—primarily nurses) in the sepsis management scenario.

1. N/A

2. N/A

3. N/A

9.3.4   STRESS

All of the above emergent factors can result in stress. In addition, there can be physical/environmental stressors (e.g., crowdedness, heat, vibration, noise, air quality, prolonged physical effort, etc.), and psychological stressors (e.g., perceived threats to one's person, suddenness of events, mental, or emotional state and condition). The stress one experiences is the responses one has to the various stressing factors. Stress can result in deteriorated performance, but also physical (e.g., headaches), mental (e.g., worrying), and emotional (e.g., depression) harm.

Use the detailed checklist to recognize emergent stress.

1. Are there environmenta—physical and/or human/organizational—stressors?

2. Do these stressors push the individual or team beyond their capabilities?

3. What are the stress symptoms?

4. Do the people handle stress well?

The following answers assess the fatigue and loss of vigilance of the people (the relevant unit(s) of analysis—primarily nurses) in the sepsis management scenario.

1. Since sepsis management, within the scenario, takes place in the ED, there are many stressors typical to the ED environment such as noise, clutter, over-crowding, too many patients, too many tasks.

2. The individual or team engaged in sepsis management can be pushed beyond their capabilities, given the presence of the stressors common to the ED environment.

3. In this case, the stress signs included a lack of patience, one nurse left the premises for 10 min. Another nurse cried.

4. The nurses' ability to handle the situation varied among team members. The triage nurse was quite discouraged with the queue of patients he still had to interview and examine.

9.3.5   STRESS, FATIGUE, AND LOSS OF VIGILANCE—A SYNTHESIS

The emergent factors stress, fatigue, and loss of vigilance have to do with too little or too much workload. What those factors tell us is that too little or too much workload can result in degraded performance. Is there such a thing as not too little and not too much workload, and how is that related to performance?

The answer is: Yes, there is, and it does relate to performance. It is based on the seminal and interesting findings on the relations between task difficulty, level of arousal, and performance (Yerkes and Dodson, 1908).

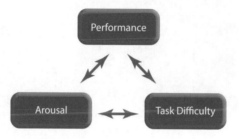

Figure 9.2: Task difficulty and arousal affect performance.

The original findings suggest that with easier tasks, performance improves with arousal level and then remains good regardless of how high the arousal level gets. However, with more difficult tasks, too little or too much arousal is associated with poorer performance. The interesting and important part is that there seems to be a level of arousal, not too high and not too low, that is associated with optimal performance. Typically, stress level is associated with arousal level. The relation between arousal level or stress and performance resembles an inverted U-shaped curve, as can be seen in Figure 9.3.

When you recognize stress as an emergent factor, try considering the level of stress, its relation to possible loss of vigilance (too little stress) and fatigue (too much stress and workload), and how those influence performance.

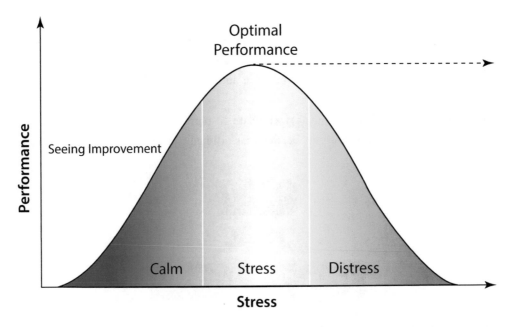

Figure 9.3: A simplified inverted U-shaped curve representing the relation between arousal level or stress and performance.

9.4 SUMMARY OF EMERGENT FACTORS

We are at the third interim summery in the MARC process. Assess the severity of each emergent factor in the same manner you assessed the severity of fit problems (see Table 8.1). We propose using Table 9.1 for that summary.

Table 9.1: Template for the summary table of the emergent factors and their severity, along with the mapped problems and their severity

Key Factors	Problem area	Problems	Severity	Emergent Factors	Severity
Goals and Tasks	Goals				
	Tasks				
Physical Environment	Physical space				
	Layout of sub-spaces and general locations				
	Artifact and device locations and access				
	Ambient conditions				
	Usability of devices and tools				
Human Environment	Teams and groups				
	Organizations and organizational culture				
	Rules, regulations, and policies				

Table 9.2 is an example of the summary of severity of emergent factors in the sepsis management scenario.

Key Factors	Problem area	Problems	Severity	Emergent Factors	Severity
	Table 9.2: Summary table of the emergent factors in the sepsis management example and their severity, along with the mapped problems and their severity				
Goals and Tasks	Goals				
	Tasks	Variable level of skill among nurses in recognizing that a patient is septic: Fairly common; when present likely to be a problem with major impact on outcomes	High	Increased mental workload	High
Physical Environment	Physical space	Crowdedness	Medium	Some stress	Medium
	Layout of sub-spaces and general locations	Challenges to providers' ability to observe and monitor a septic patient	Medium	Some discomfort and frustration; some stress	Medium
	Artifact and device locations and access	Sepsis workbook not present: Uncommon; when missing, can be accessed somewhere in the ED; good clinical skills can negate the need for it.	Low	Increased mental workload; some discomfort and frustration; little stress	Medium
		Lack of readily available information and alerts regarding abnormal physical and lab-based findings: Common, substantial impact on outcomes	High	High stress	Medium
	Ambient conditions	Noise	Low	Increase mental workload; little stress	High
	Usability of devices and tools	Inconsistent and hard to reach information	Medium	Increase mental workload; some stress	Medium

Human Environment	Teams and groups	Once sepsis is recognized, team members communicate well regarding sick patients. However, failing to close communication loops may affect, likely commonly, and with substantial consequences	High	Increase in dissatisfaction and frustration	Medium
	Organizations and organizational culture	Poor capability to promote learning in the organization can have major consequences	High	Increase mental workload; high stress	Medium
	Rules, regulations, and policies	Problem with adopting and adhering to policies and regulations	Low	Little stress	Medium

Note the severity ratings of the problems and the emergent factors in the sepsis management scenario. They are not identical. There could be factors whose severity will be rated as low (e.g., noise), whereas the emergent human factors associated with this environmental factor (e.g., increased stress and mental workload) could be rated high. In the next and almost final step, identifying the most influential factors, we will consider both ratings.

9.5   INTERIM BRIEF #2

By completing the "Recognize" step in the MARC process, we covered the two bottom tiers of the human factors framework. Now it is time to summarize what we have so far in our second brief.

Brief Requirements: Use your assessment of the fit between people's capabilities, their task demands, and the environment by answering the following:

- In what ways do the task requirements fit the people's capability to perform them? What gaps did you identify?

- In what way does the physical environment support people's capability to perform their tasks? How does the physical environment inhibit people's capabilities?

- In what way do the tools and devices support people's capabilities to perform their tasks? In what way do they restrict their capabilities?

- In what way does the human environment support people's capability to perform their tasks? How does the human environment inhibit people's ability to perform?

- Given your above assessment of the fit, what are the problems and what is their severity?

- Given your above assessment of the fit, what are the emergent factors from the combination of tasks and environment, on the one hand, and human capabilities, on the other?

- Summarize and present your fit assessment and emergent factors in a brief.

Your brief should, as a minimum, include information about:

- Background: reminder of the mapping.

- Tasks' fit: briefly outline issues of good and poor fit of tasks with people's capabilities.

- Physical environment's fit: briefly outline issues of poor fit of physical environment with people's capabilities to perform tasks.

- Tools and devices fit: briefly outline usability issues (poor fit of using tools and devices with people's capabilities to perform tasks).

- Human environment fit: briefly outline issues of poor fit of human environment with people's capabilities to perform tasks.

- Summary of problems ordered by your assessment of their severity.

- Emergent factors: briefly outline the emergent factors you recognize.

CHAPTER 10

Conclude: Performance and Outcomes

The Conclude step of the MARC process looks at the following two "simple questions.

1. Can we explain what happened, or, can we explain what may happen?

2. What should we do about it?

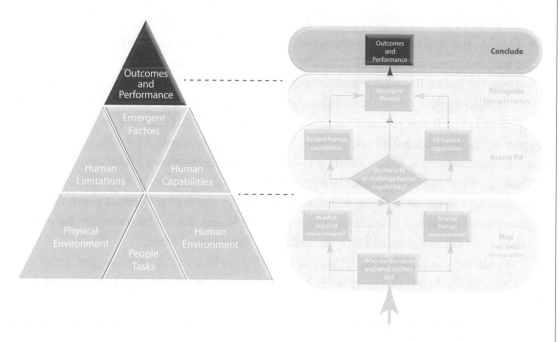

Figure 10.1: The Conclude step.

We answer these questions in the Conclude step by looking for the most influential factors, the people's performance, and the outcomes.

10.1 THE MOST INFLUENTIAL FACTORS

We define the most influential factors in terms of their validity and relevance to the objectives of the analysis. With that in mind, we can begin to identify them.

10.1.1   SCOPE ACCORDING TO VALIDITY AND RELEVANCE

Recall that the goal of performing the human factors analysis is to propose *valid* interventions and mitigations *relevant* to the objectives of the analysis we have defined. In order to propose those valid and relevant interventions and mitigations, we must first determine what is "relevant" and what is "valid." Relevant and valid interventions and mitigations are the ones that address the factors—environmental and human—that were found to be associated with problems, particularly severe problems that are likely to influence performance and outcomes. In other words, we should identify those factors that are most influential and propose interventions and mitigations that address them. By scoping the interventions and mitigations accordingly, we increase the chance of achieving those objectives.

10.1.2   IDENTIFY THE MOST INFLUENTIAL FACTORS

We are often tempted to focus on the factors that are likely to degrade performance and result in harmful outcomes, and mitigate them. But at the same time, we want to preserve and even strengthen the factors that have a positive and facilitating impact on performance and outcomes. Thus, identifying which are the most influential factors that you want to address has to do with the goals and objectives of your analysis. If the goal of your analysis is proactive: improve a process, support the implementation of new systems or devices or procedures, or help design a new site, then the most influential factors should also include those factors that are likely to support and facilitate good performance and positive outcomes. If the goal of your analysis is to investigate and understand an adverse event, the most influential factors are the ones that reflect problems with a high severity rating, as well as resulting in poor emergent factors. But in both cases, we always want to attend to all influential factors; it is just a matter of focus and priorities.

To identify the most influential factors, we revisit the issues of fit and emergent factors and their severities. The idea is that there is a link between issues of fit and emergent factors as we have seen before.

Combine the issues of fit and the emergent factors to determine the most influential factors as is shown in Table 10.1.

Table 10.1: Template of summary table of the most influential factors

Key Factors	Problem area	Problems	Severity	Emergent Factors	Severity	Most Influential Factors
Goals and Tasks	Goals					
	Tasks					
Physical Environment	Physical space					
	Layout of sub-spaces and general locations					
	Artifact and device locations and access					
	Ambient conditions					
	Usability of devices and tools					
Human Environment	Teams and groups					
	Organizations and organizational culture					
	Rules, regulations, and policies					

Here are the influential factors determined in the sepsis management scenario. Note that factors that did not get identical severity ratings for the problems and emergent factors, the relevant factor was nevertheless considered an influential factor even if only one of the ratings was high.

Table 10.2: Influential factors in the sepsis management scenario						
Key Factors	**Problem area**	**Problems**	**Severity**	**Emergent Factors**	**Severity**	**Most Influential Factors**
Goals and Tasks	Goals					
	Tasks	Variable level of skill among nurses in recognizing that a patient is septic: Fairly common; when present likely to be a problem with major impact on outcomes	High	Increased mental workload	High	Insufficient experience in sepsis management
	Physical space	Crowdedness	Medium	Some stress	Medium	Crowdedness
	Layout of sub-spaces and general locations	Challenges to providers' ability to observe and monitor a septic patient	Medium	Some discomfort and frustration; some stress	Medium	Layout - insufficient support of task
Physical Environment	Artifact and device locations and access	Sepsis workbook not present: Uncommon; when missing, can be accessed somewhere in the ED; good clinical skills can negate the need for it.	Low	Increased mental workload; Some discomfort and frustration; some stress	Medium	Inappropriate location of sepsis workbook
		Lack of readily available information, and alerts regarding, abnormal physical and lab-based findings: Common, substantial impact on outcomes.	High	Some stress	Medium	Inappropriate location of other information

Physical Environment	Ambient conditions	Noise	Low	Increase mental workload; some stress	High	Noise
	Usability of devices and tools	Inconsistent and hard to reach information	Medium	Increase mental workload; some stress	Medium	Poor usability of EHR in support of the task
Human Environment	Teams and groups	Once sepsis is recognized, team members communicate well regarding sick patients. However, failing to close communication loops may affect, likely commonly, and with substantial consequences	High	Increase in dissatisfaction and frustration	Medium	Ineffective communication
	Organizations and organizational culture	Poor capability to promote learning in the organization can have major consequences	High	Increase mental workload; some stress	Medium	Insufficient learning and professional development
	Rules, regulations, and policies	Problem with adopting and adhering to policies and regulations	Low	Emergent Factors	Medium	Ineffective policy implementation

10.2 PERFORMANCE AND OUTCOMES: HOW ARE THEY DIFFERENT?

Before we delve into further conclusions about performance and outcomes, let us first discuss the similarities and differences between these concepts. Performance refers to some measure of how well or poorly the task was executed. Outcome refers to the consequences of executing the task in a certain way. If it is that simple, why do we bother discussing this? Because sometimes it is hard to distinguish between them. Committing an error is the best example of the potential vague distinction between performance and outcome. Is making an error a performance metric or is it the outcome? Let us be more specific with a hypothetical example. One of the unfortunate frequent medical errors is wrong medication dosage. If we did our human factors analysis we could have discovered some factors that brought about this error. Is a wrong medication dose a performance metric or an outcome? In this book, we make a distinction between the two concepts, because typically there are further consequences to performance. In our example of making an error with the dosage, there are further consequences to errors. Let's explore the distinction between performance and outcomes further.

10.2.1   PERFORMANCE: EFFECTIVENESS, HUMAN ERROR, AND EFFICIENCY

A critical part in the conclusion of the analysis is to determine how the fit issues and emergent factors influence performance. Performance is defined by the unit of analysis and how we measure it. With respect to the unit of analysis, we determine if we are talking about the performance of an individual, a team, an organizational unit, or a whole system. In order to simplify the analysis, we propose here a small set of metrics that can be relevant to the varying graininess of units of analysis. The metrics we propose are effectiveness, human error, and efficiency.

Effectiveness

Effectiveness refers to the extent of success or failure in completing the task. A task could be at the level of the clinical task itself (e.g., administer medication), to the more detailed actions required to accomplish the task (e.g., press on a button in the IV pump). That would also include all cognitive tasks, including situation assessment, communication and coordination, etc.

Human Error

Inherent in the effectiveness of performance are *human errors*. A human error is one's failure to execute a desired or required action/task in the relevant context (time and place). Failure could be omitting the desired/required action, or committing it in the wrong way (e.g., incomplete, wrong timing, wrong place, incorrect direction or quantity, etc.). The working assumption here is that such failures are associated with problems with any of the factors discussed up to now (e.g., tasks too complex or too stressing; physical environment inappropriate of creating many interruptions and distractions; human environment failures; task and environment do not fit human capabilities; etc.). When looking at the broader context of human error, we should consider Reason's Swiss Cheese Model for the "trajectory" of an error due to latent conditions through holes in defenses all the way to the active failures (Reason, 1990, 2000).

A simple taxonomy of human errors was originally proposed by Norman (1988). The basic idea in this taxonomy is the division of errors into two main types: mistakes and slips. Mistakes are typically errors of perception, attention, and cognition. Those errors could include failures with detection, recognition, or identification of information (e.g., failing to notice a low oxygenation level). They could include failure with directing attention or dividing attention when required (e.g., failing to attend to both vital signs and patient's skin color). They could include failure with comprehension and assessment of information (e.g., failing to determine that the observed symptoms reflect deterioration in patient's status). And finally, they could include failure in intentions and plans for action. Slips are failures in action and executions of intentions and plans.

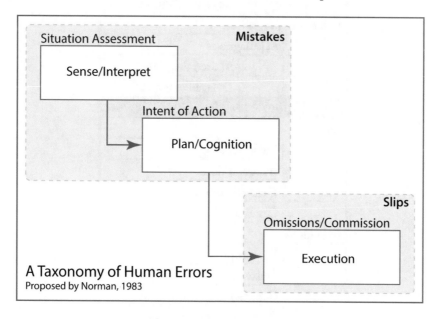

Figure 10.2: A taxonomy of human errors proposed by Norman (1988).

To get a more concrete idea about types of errors in healthcare, here is an example of some likely human errors (taken from Elder and Dovey, 2002, p. 930).

Table 10.3: Likely human healthcare errors		
Diagnosis		
Related to symptoms		
	Misdiagnosis	
		Missed diagnosis
		Delayed diagnosis
Related to prevention		
	Misdiagnosis	
		Missed diagnosis
		Delayed diagnosis
Treatment		
Drug		
	Incorrect drug	
	Incorrect dose	
	Delayed administration	
	Omitted administration	
Non-drub		
	Inappropriate	
	Delayed	
	Omitted	
	Procedural complication	
Preventive services		
Inappropriate		
Delayed		
Omitted		
Procedural complication		

Efficiency

The extent of resources expended in completing the task. The most important resource we want to consider with respect to efficiency is time. That is, efficient performance is one where the person, team, or unit successfully completes the task in an acceptable time according to some established or accepted benchmark.

Summarize the performance associated with each of the most influential factors in the following table.

Key Factors	Problem area	Most Influential Factors	Performance
Goals and Tasks	Goals		
	Tasks		
Physical Environment	Physical space		
	Layout of sub-spaces and general locations		
	Artifact and device locations and access		
	Ambient conditions		
	Usability of devices and tools		
Human Environment	Teams and groups		
	Organizations and organizational culture		
	Rules, regulations, and policies		

Table 10.4: Performance associated with the most influential factors

Table 10.4 summarizes the performance associated with each of the most influential factors in the sepsis management example.

Table 10.4: Performance associated with each of the most influential factors for the sepsis management scenario

Key Factors	Problem area	Most Influential Factors	Performance
Goals and Tasks	Goals		
	Tasks	Insufficient experience in sepsis management	Likely errors degrading effectiveness and causing delays
Physical Environment	Physical space	Crowdedness	
	Layout of sub-spaces and general locations	Layout insufficient support of task	Delays
	Artifact and device locations and access	Inappropriate location of sepsis workbook; Inappropriate location of other information	Likely errors degrading effectiveness and causing delays
	Ambient conditions	Noise	Degrading effectiveness
	Usability of devices and tools	Poor usability of EHR in support of the task	Degrading effectiveness and causing delays
Human Environment	Teams and groups	Ineffective communication	Likely errors degrading effectiveness and causing delays
	Organizations and organizational culture	Insufficient learning and professional development	Degrading effectiveness
	Rules, regulations, and policies	Ineffective policy implementation	Degrading effectiveness

10.2.2 OUTCOMES: FACTUAL, LIKELY, AND DESIRED

Before we discuss outcome categories that would lead us to proposing interventions and mitigations, we should remind ourselves of the distinction between the two different analysis goals. If we are in a proactive context, the analysis helps us determine likely outcomes of performance as a function of likely problems and emergent factors. However, if we are in a retrospective context investigating an adverse event, we know what the performance and the outcomes were, and in most cases those would be undesirable outcomes. In both cases we need to go beyond the factual outcomes or the likely outcomes, and aim to propose interventions and mitigations that will bring about desired outcomes.

Outcomes can be categorized as follows:

- patient-specific,

- healthcare professional,

- other people (e.g., family), or

- organizational.

Below are questions to consider for each category.

Patient-specific Outcomes

- How did the performance of the unit of analysis affect the patient's health (clinical outcomes)?

- Was the patient satisfied with the performance?

- What were the monetary costs involved?

Healthcare Professional Outcomes

- What was the impact of performance on the healthcare professional?

- What emotional impact did they experience?

- Are healthcare professionals satisfied with their work?

Other People Outcomes

- Were the other people satisfied with the performance?

- What monetary costs were involved?

Organizational Outcomes

- What operational impacts did the performance have, such as wait times, discharge times, and hospitalization times?

- What monetary costs were involved?

- What legal consequences were there?

- How did performance influence the organization's reputation?

Summarize the outcomes in relation to the influential factors and performance in Table 10.5. Note that the column for the outcomes is not divided into lines corresponding to the specific factor categories. This is to signify that the outcomes could be related to several influential factors. In other words, outcomes could result from the inter-play between several factors.

Table 10.5: Summary of outcomes

Key Factors	Problem area	Influential Factors	Performance	Outcomes
Goals and Tasks	Goals			
	Tasks			
Physical Environment	Physical space			
	Layout of sub-spaces and general locations			
	Artifact and device locations and access			
	Ambient conditions			
	Usability of devices and tools			
Human Environment	Teams and groups			
	Organizations and organizational culture			
	Rules, regulations, and policies			

Table 10.6 is the summary of outcomes for the sepsis management scenario.

Table 10.6: Summary of outcomes for the sepsis management scenario

Key Factors	Problem area	Influential Factors	Performance	Outcomes
Goals and Tasks	Goals			
	Tasks	Insufficient experience in sepsis management	Likely errors degrading effectiveness and causing delays	Worsening physiology and organ dysfunction
Physical Environment	Physical space	Crowdedness		Need for more aggressive measures
	Layout of sub-spaces and general locations	Layout insufficient support of task	Delays	Longer recovery time and hospital admission
	Artifact and device locations and access	Inappropriate location of sepsis workbook; Inappropriate location of other information	Likely errors degrading effectiveness and causing delays	Increased morbidity and mortality Poor patient and caregiver experience and satisfaction, loss of trust in the care team, frustration, and financial losses
	Ambient conditions	Noise	Degrading effectiveness	
	Usability of devices and tools	Poor usability of EHR in support of the task	Degrading effectiveness and causing delays	Increased provider workload and stress, potential guilt, moral distress, and burnout
Human Environment	Teams and groups	Ineffective communication	Likely errors degrading effectiveness and causing delays	Increased wait times in the ED, preventable ICU admissions, and longer ICU and hospital stays
	Organizations and organizational culture	Insufficient learning and professional development	Degrading effectiveness	
	Rules, regulations, and policies	Ineffective policy implementation	Degrading effectiveness	Increased cost to the hospital and healthcare system Risk of litigation and damaged reputation for the hospital

CHAPTER 11

Interventions and Mitigations

The purpose of performing the human factors analysis is to improve performance and outcomes through interventions and mitigations. Regardless of whether we are investigating an adverse event, contributing to a process or a site improvement, or designing a new system implementation, our goal is to offer improvements. A challenge in offering improvements is that there are not always enough resources (for example, time and money) to implement all of the recommendations. Moreover, with many stakeholders involved, an improvement for one stakeholder may conflict with the interests of another one.

To accommodate the process of identifying possible improvements and then choosing which ones to adopt, we offer a number of tools.

- Fit the granularity of the recommendations to the stage in the decision-making process.

- Systematically address all intervention and mitigation categories.

- Involve all stakeholders in a quantitative tradeoff analysis in order to prioritize.

- Share the brief with the stakeholders and use it as a living document, building on feedback.

11.1 GRANULARITY OF THE RECOMMENDATIONS

Fitting the granularity of the recommendations to the stage in the decision-making process can improve efficiency. Consider having the mitigation proposal on three levels.

1. Strategic Goals: Express the high-level goals of the proposed intervention and mitigation.

2. Objectives: Define the objectives derived for a given strategic goal.

3. Specifications: Specify the most concrete requirements and instructions on what to do to meet the objectives and goals.

Figure 11.1 shows an example of the three granularity levels.

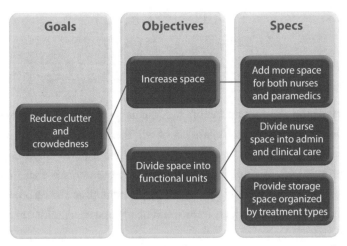

Figure 11.1: An example of three granularity levels of proposing interventions and mitigations.

You could provide your recommendations on any of these levels. By starting at the high level and then going into detail, you can discuss the issues in general with the stakeholders and only delve into the more detailed recommendations based on their feedback. This can contribute to the efficiency of the process. Involving stakeholders in this process can also have a positive effect on their acceptance of the recommendations, although the consultation process runs the risk of adding discussion time to the process as well.

11.2 INTERVENTION AND MITIGATION STRATEGIC GOALS

Determine the mitigation strategy you would propose for the most influential factors in each of the main categories from the foundation tier in the human factors conceptual framework. These are based on the answers to the questions in the Assess the Fit and Recognize Emergent Factors phases (Section 3.4.2) of the HF-MARC process, and the prioritized problem inventory you compiled. Table 11.1 offers some examples of mitigations strategies for each factor category for you to consider.

Table 11.1: Examples of mitigation strategies

Key Factors	Factor Category	Mitigation Strategy
Goals and Tasks	Goals	Re-visit goals; clarify goals; teach and advocate the goals.
	Tasks	Re-design tasks (easier, less complex, less multi-tasking, etc.); reduce the number of tasks per individual; re-allocate and re-distribute tasks; revise workflows; add aids and technologies.
Physical Environment	Physical space	Expand the space (for example, it is possible to reduce the number of artifacts and/or people in the space); change job definitions and reallocate people in space; change work routines.
	Layout of sub-spaces and general locations	Rearrange elements; reduce the number of elements; revise tasks and workflow.
	Artifact and device locations and access	Consider changing: locations of the artifacts and people; task requirements; workflow; redesign artifacts.
	Ambient conditions	Consider changing the ambient conditions: better lighting; less noise; better temperature; follow OSHA guidelines and regulations.
	Usability of devices and tools	Re-design the device; procure a different device; improve training and support; change the ways procurement decisions are made; involve usability professionals in the design and procurement decisions; perform early usability testing.

Key Factors	Factor Category	Mitigation Strategy
Human Environment	Teams and groups	Consider team training (e.g., crisis resource management, communication, etc.); change team structure; change roles and responsibilities; add technology in support of teamwork.
	Organizations and organizational culture	Consider organizational changes (structure, management, internal communications, reward and promotion system, accountability and reporting systems, information technology).
	Rules, regulations, and policies	Develop policies. revise existing policies. make policies more accessible. communicate policies and train in following them.

Table 11.2 provides sample interventions and mitigations for each problem area and its influential factors in the sepsis management scenario.

Table 11.2: Sample interventions and mitigations for sepsis management scenario

Key Factors	Problem Area	Influential Factors	Interventions and Mitigations
Goals and Tasks	Goals		
	Tasks	Insufficient experience in sepsis management	Train staff (e-modules, small group learning, simulation).
Physical Environment	Physical space	Crowdedness	Keep working on strategies to alleviate bottlenecks and streamline patient flow through ED.
	Layout of sub-spaces and general locations	Layout insufficient support of task	As a new ED is being currently planned, engage in designing spaces that would support interaction and communication.

Key Factors	Problem Area	Influential Factors	Interventions and Mitigations
Physical Environment	Artifact and device locations and access	Inappropriate location of sepsis workbook; inappropriate location of other information	**Workbook:** Design a process to ensure that the workbook is available at all times. **Other information:** Difficult to redesign the current electronic chart. Potentially: work with the provider on introducing alerts (with a focus on usability) to providers while managing septic patients.
	Ambient conditions	Noise	Engage security when crowd noise is there. Use technology to reduce noise.
	Usability of devices and tools	Poor usability of EHR in support of the task	See Other Information, above. Familiarize staff with good design principles so that when they provide feedback during the procurement process, their feedback is more helpful.
Human Environment	Teams and groups	Ineffective communication	Train staff in effective communication (issues: funding, time).
	Organizations and organizational culture	Insufficient learning and professional development	Same as above: the organization will need to find ways, and identify required resources, to ensure learning.
	Rules, regulations, and policies	Ineffective policy implementation	Same as above: The organization will need to identify strategies to facilitate policy implementation.

At this point, we may end up with many proposals for interventions and mitigations, at various granularity levels. Do we propose all of them? Are they all feasible? Are they all of equal importance and potential impact? We propose a final analysis, a trade-off analysis, whereby you can help stakeholders prioritize the intervention and mitigation proposals.

11.3 PRIORITIZE THE INTERVENTIONS USING A TRADEOFF ANALYSIS

Trade-offs are fundamental in deciding on interventions and mitigations, and are a thorny challenge in addressing any case, site, or procedure. The healthcare community includes a variety of professionals involved as stakeholders. Each stakeholder may have different objectives, constraints, and approaches.

Making decisions involving trade-offs (for example, when fulfilling one stakeholder's objective may take away from another) is a complex and difficult task—conflicting needs may become a serious obstacle to making objective and rational decisions. There are quantitative tools to help overcome this challenge, including trade-off analysis.

Trade-off analysis provides an objective approach for considering the multiple objectives and constraints; and reaching weighted decisions. Trade-off analysis is a tool for identifying alternatives and constraint criteria, collecting the assessments and/or preferences of stakeholders relative to the alternatives and criteria, computing the combined relative weights of the trade-offs, and supporting decision-making. The method is adapted from several sources (Hagen, 1967; Meister, 1985; Pugh, 1991; Otto and Wood, 2003) and simplified here to provide a pragmatic and lightweight analysis.

Math Alert! There is a bit of math involved in running a tradeoff analysis. We provide the relevant equations for each step of the analysis. However, you need not be too concerned with those equations, and you can simply follow these main steps that include basic algebraic operations (summarized in Table 11.3).

1. Determine criteria for trade-off decision: identify and define the key objectives and constraints introduced by the various stakeholders. These could be any of the following:

   a. Implementation feasibility: The possibility and ability to implement the intervention.

   b. Immediate costs: The cost of implementing the intervention.

   c. Life-cycle costs: The longer-term costs of sustaining the intervention.

   d. Viability: The likelihood of the intervention to succeed and improve outcomes.

   e. Sustainability: The sustainability of the impact of the intervention for a longer-term.

   f. Acceptability: The likely acceptability of the intervention by the organization.

2. Determine relative weight for each decision criterion: Rank the position of a given criterion relative to the other criteria (for example, overall the viability of the intervention is the most important). Then compute the relative weight of that criterion by dividing a given rank by the total of all the ranks.

3. List the most influencing factors that represent the key problems you want to mitigate.

4. Determine the tradeoff score (relative weight) of each problem by criterion: Rank the position of each problem for a given criterion. For example, the problem of cluttered workspace would rank high on the mitigation feasibility, but relatively lower on the mitigation impact likelihood. Then compute the relative weight of a given problem for a given criterion (by dividing it by the sum of all ranks for that given problem).

5. Compute the overall tradeoff score for each problem: compute the overall tradeoff score of a given problem by summing up all its relative scores for all the criteria. The overall tradeoff score for a given problems is its relative weight in consideration of all the criteria.

The entire process, along with the relevant equations for each step, is summed up in the following matrix.

Table 11.3: A generic tradeoff analysis matrix along with the math equations for the computations

		Decision Criteria ①				
		Criterion 1	Criterion 2	Criterion 3	Criterion n	
	Rank	R_i ②				
	Relative	W_i				
Problems ③	Problem 1	S_{ij} T_j ④				OT_j ⑤
	Problem 2					
	Problem 3					
	Problem n					

For step 2: $W_i = \dfrac{R_i}{\sum_{i=1}^{N} R_i}$ whereby R_i is the rank position for criterion i, and W_i is the relative weight for criterion i

For step 4: $T_j = W_i * S_{ij}$ whereby S_{ij} is the rank of problem j for criterion i, and T_j is the tradeoff score (relative weight) of problem j for criterion i

For step 5: $OT_j = \sum_{i=1}^{N} W_i * S_{ij}$ whereby OT_j is the overall tradeoff score for problem j

Table 11.4 is an example of the tradeoff analysis for the sepsis management scenario.

Table 11.4: Tradeoff analysis for the sepsis management scenario (each ranking is on a scale from 1–5, with 1 being very low, and 5 being very high)

		Decision Criteria			
		Mitigation Feasibility	Mitigation Cost[4]	Mitigation Time[4]	Mitigation success liklihood	
	Rank	3	2	2	4	11
	Relative	.27	.18	.18	.36	
Problems	Insufficient experience	4 1.08	5 (1) .18	5 (1) .18	5 .18	3.24
	Inappropriate location of workbook	5 1.35	2 (4) .72	2 (4) .72	5 1.8	4.59
	Ineffective communication	5 1.35	4 (2) .36	4 (2) .36	5 1.8	3.87
	Insufficient learning and development	3 .81	5 (1) .18	5 (1) .18	4 1.44	2.61

[4] Reverse scale (original rank in brackets).

Conclusions: The inappropriate location of the workbook is the problem that has the highest score in the tradeoff analysis with respect to implementing the relevant intervention. This is followed by addressing the ineffective communication problem.

11.4 FINAL BRIEF #3

Brief Requirements: This brief is the culminating brief for your analysis. Here are some guidelines for preparing it.

1. Use the key problems you have identified in your fit assessment, and the emergent factors you have recognized, and answer the following.

    a. In what ways might the fit (or poor fit or lack of fit) influence performance?

    b. In what way do the emergent factors influence or could influence performance?

    c. What are the most influencing factors?

    d. In what way do the most influencing factors influence performance?

    e. In what way do the most influencing factors influence outcomes?

2. Given your above conclusions, what are your recommended and prioritized mitigations?

Your brief should, as a minimum, include information about the following:

1. Background: reminder of the context, the key problems by their severity, and the emergent factors. Note, this is an important part since members of the committee did not hear the preceding briefs.

2. Implications of the problems and emergent factors to performance of people and their tasks.

3. The most influencing factors.

4. Implications to performance and care outcomes.

5. Recommended mitigations by priorities.

CHAPTER 12

This is Not the End

Congratulations on reaching the end of this book! We actually see this as a beginning. As you use the book, note which strategies are especially useful to you. Consider revisiting others as you gain experience.

As we have suggested at the beginning, the approach presented in this book aligns with other quality improvement frameworks (such as that of the Agency for Healthcare Research and Quality, 2013, which includes the Plan-Do-Study-Act Model for Improvement; Langley et al., 1996) and can help you **consider the fuller picture** and ultimately achieve better results.

We welcome your feedback. What worked for you? Where could the explanations benefit from more detail? More examples? Anything else?

How does this differ from other strategies you use? How is it similar?

In the spirit of this book, we are committed to continual quality improvement. Your feedback can help us help you in the next version.

Appendix A: The Detailed Checklist

MAP THE CONTEXT

People

1. Who are the people involved directly?

2. Are there additional relevant stake-holders? If yes, who are they?

3. What are their typical characteristics with respect to:

   - roles,

   - profession, qualification,

   - knowledge and skills,

   - experience,

   - context of work or use,

   - role in the analyzed situation or context, and

   - level of involvement and responsibility.

Missions, Goals, Tasks

1. What is or are the goals of each of the individuals and/or the team? Examples include administration, triage, diagnose, treat, and others.

2. What are the physical, overt tasks the individual or team performs? Examples include inserting an IV and administering a medication.

3. What are the physical activities required for performing the task? Examples include reaching, grabbing, walking, pointing, and inserting.

4. What are the perceptual, attentional, and cognitive tasks the individual or team performs to achieve this? Examples include monitoring relevant elements of the

environment by looking, listening, and touching, noticing changes, and evaluating the meanings associated with these. It includes situational awareness, which reflects how each person understands what is happening with respect to the mission or task being performed.

5. What are the inter-relations between the tasks? Note, this part of the mapping is addressed in a more elaborate fashion in the section on Task Analysis—Map the Inter-relations between the Tasks (Section 6.2.1).

Physical Environment

Space and Layout

1. What space is available?

2. Is the space divided into sub-spaces? If yes, what are the criteria for this division?

3. How are the sub-spaces arranged spatially?

4. Are there access paths and routes?

5. Where are the entry and exit points?

Artifact and Device Locations and Access

1. Where are all the required elements (people, displays, anything) located?

2. Where are the artifacts used by individuals to perform tasks located?

3. Where are the artifacts required by several people to perform their tasks located?

4. Where are the people in the space located relative to the location of artifacts, devices, and information they require to perform their tasks?

Ambient Conditions

1. What light is available to perform the task? What are the light levels? Are there reflections anywhere in the workspace?

2. Is there noise? What is the nature of the noise? What is the noise level?

3. What is the temperature in the work environment?

4. What is the air quality in the work environment?

5. Are there vibrations in the work environment? If yes, does the frequency or intensity of vibrations create disturbance?

6. Are there any usual or unusual odors in the work environment?

Display

1. What is the subject matter (content) of the information? (For example, vital signs, HR, BP, etc.)

2. What are the information characteristics? (For example, high-low, exact number, trend, etc.)

3. What is the level of detail in the information? (For example, do we need to show very high detail or enough to give ranges? Is a very accurate task required or qualitative info is enough?)

4. What is the rate of update of the information?

5. What is the level of tolerance for errors in the information?

6. In what ways or modalities is the information conveyed?

7. In how many ways is the information conveyed?

8. How is the information presented and conveyed? (For example, visually: text vs. graphics, numeric vs. analog, size, color; auditorily: sound vs. speech, amplitude, frequency, etc.)

9. Does the device include a haptic, tactile display? If yes, what is the manner of the display: amplitude, frequency, temperature?

10. What are the contextual considerations in the way the information is conveyed? (For example, indoors vs. outdoors, lighting conditions and reflections, movement, etc.)

Controls

1. What is the controlled function or action? (For example, raise or lower threshold, etc.)

2. What is the human modality for the control? (For example, hand, finger, foot, speech, etc.)

3. What are the control actions? (For example, push, pull, press, tap, rotate, speak, look, move, think, etc.)

Dialogue

1. How do users start the dialogue, the task?

2. Is the dialogue continuous or in discreet steps?

3. Can the user undo or reverse steps?

4. What is the style of the dialogue? (For example, selecting commands from menus or keying numbers based on a voice menu? Direct manipulation?)

5. How do users complete or abort the dialogue?

Human Environment

1. Who are the other people in the workspace (healthcare professionals, patients and families, any other personnel)?

2. What do they do?

3. Are they members of a group or a team with these other people? Are the key stakeholders members of a group or a team with these other people?

4. What is the structure of the team (flat vs. hierarchical)?

5. Is there leadership? Who is/are the leaders? How did they become leaders?

6. What do team members know about each other?

7. How do team members communicate? How often? What about?

8. How do team members coordinate and collaborate?

9. What is the definition of roles?

10. What is the allocation of workload?

11. What are the expectations and understanding of everyone's role and what is supposed to be done and when (mental models)?

Organizations and Organizational Culture

1. What is the division and definition of roles within the organization?

2. What is the organizational structure?

3. What are the staffing practices and policies?

4. What are the organization's approaches toward workers, safety, quality, and accountability?

5. What kind of learning takes place in the organization?

6. How is information transferred within the organization?

7. What are the development, training, and certification opportunities?

Rules, Regulations, and Policies

1. What are the available policies?

2. How are the policies designed and communicated?

3. Are there too many workarounds? If yes, what is the reason?

4. How is workload allocated and managed? Consider work hours, shifts, rotations, rest periods, night work.

5. What is the practice in terms of sticking to rules and policies?

ASSESS FIT

People and Tasks

The People Perspective

1. Do the people have the relevant training, certification, and experience to perform the tasks?

2. Is the person or are the people in the team motivated to do their job?

3. Do they have positive attitudes toward safety and quality?

4. Are the people aware of possible risks and hazards in their work place?

5. Do the people tend to avoid taking unnecessary risks while performing their tasks?

6. Do the people have a perception that they are in control of what is happening?

7. How do the people handle stress?

8. Are people high-strung or laid-back?

9. Can people divide their attention between other people and their tasks?

10. How do people make the decisions required to accomplish their tasks? Do they make the decisions effectively?

11. Do people seem to exhibit any cognitive biases?

The Task Perspective

1. Are there too many tasks for the individual or team?

2. Are they required to perform too many tasks simultaneously?

3. Is the time available to perform tasks too short?

4. Are the tasks too complex? Too difficult? Too easy?

5. Is there too much information?

6. Is the information too complex to receive, retain, or use while performing tasks?

7. Can people divide their attention between several tasks?

8. Can the user track or monitor information for a prolonged period?

9. Can people make decisions given the time they have, the number of tasks they need to deal with, and/or the amount of information they need?

10. Is the required physical effort appropriate to the people?

11. Do people exert physical effort for prolonged periods?

12. Do people exert physical effort that challenges the respiratory or cardio-vascular system?

Physical Environment

Space and Layout

1. How well does the space fit the number of individuals or groups/teams?

2. Is there sufficient space for the individual or group/team to perform tasks and achieve goals?

3. Are the sub-spaces configured in a way that supports the workflow to be effective, efficient, and safe?

4. Are the access paths and routes convenient?

5. Are the entry and exit points located where they support the workflow?

6. Are the locations of doors and windows or any other fixed elements appropriate?

Location of Artifacts and Devices

1. Are the artifacts one uses for performing tasks located appropriately for purposes of reaching, viewing, and/or hearing?

2. Are the artifacts located appropriately for several people to perform their tasks?

3. Does the layout support people in focusing on information or elements they require for their tasks?

4. Can people notice and receive from their location relevant information even though they are not focusing on it?

5. Can people divide their attention between several sources of information?

6. Is information displayed visually in a way that supports reading and identification of details?

7. Are all required information and artifacts within the field of view of the people who need it? If not, does it require much eye and head movement to perceive it?

Ambient Conditions

1. Is there sufficient light to perform the task? Too much? Too little? Are there too many reflections? Are the reflections disturbing?

2. Are the lighting conditions such that they do not require adaptation to low-level lighting?

3. Is the level of the noise reasonable?

4. Does the noise create distractions and interruptions?

5. Can people hear all the auditory information they require?

6. Can they distinguish between sounds (for example, alarms) and speech?

7. Is speech intelligible?

8. Are they required to localize the source of sounds or speech, and can they do it?

9. Is the temperature in the work environment appropriate? Not too cold? Not too hot?

10. Is the air quality in the work environment appropriate?

11. Are there vibrations in the work environment? If yes, does the frequency or intensity of vibrations create disturbance?

12. Do the odors create a problem (discomfort, breathing difficulty, tearing, and any other responses)?

Device Usability

The following is a simple checklist covering the key aspects in the user interface of a device, following the ten usability heuristics suggested by Nielsen (1993).

Statement	Strongly Disagree			Undecided			Strongly Agree
	(1)	(2)	(3)	(4)	(5)	(6)	(7)
1. The work flow has a clear beginning and end.							
2. All functions are grouped to support the work flow.							
3. The device keeps the user informed about what is going on.							
4. The information appears in natural and logical order.							
5. The device allows the user to have full control.							
6. The device prevents errors from occurring.							
7. The device provides clear ways to undo actions and recover from errors.							
8. The device "speaks" the user's language.							
9. The device looks aesthetically pleasing.							
10. The device has a simple visual design.							

Human Environment

Groups and Teams

1. Does the presence or absence of any of the people in the workspace support or disrupt the workflow and tasks?

2. Can people divide their attention between other people and their tasks?

3. Does the structure of the team support teamwork and team tasks and goals?

4. Is there a leader? Is the leadership effective?

5. Do all team members share recognition/familiarity with the professional expertise or authority of each member in the team?

6. Do team members communicate effectively?

7. Do they share mental models?

8. Is there sharing of awareness and understanding of the situation (situational awareness)?

9. Does communication become disruptive (does it become an interruption or distraction)?

10. Is there effective coordination and collaboration?

11. Is there a clear definition of roles?

12. Is there a clear and fair allocation of workload?

13. Do people have good structured knowledge of who does what, when, and how (mental models)?

14. Do team members share similar expectations and understanding of everyone's role and what is supposed to be done, when, and how?

15. Are people capable of being aware of what is going on? Do they share this knowledge?

16. Do team members effectively share awareness and understanding of the situation?

Organizations, Climates, and Cultures

1. Is the division and definition of roles within the organization clear?

2. Does the organizational structure support the achievement of its goals?

3. Is the staffing sufficient or appropriate?

4. Is the organization committed to its members in a way that promotes positive attitudes and behaviors?

5. Are members of the organization involved and engaged?

6. Does the organization have positive culture with respect to safety and quality?

7. Does the organization have an accountability culture that is positive, supportive, and constructive?

8. Does learning place in the organization in a way that development and changes are supported?

9. Is there an effective transfer of information within the organization?

10. Is available training and certification sufficient or appropriate for people to perform their tasks?

Rules, Regulations, and Policies

1. Are there sufficient or appropriate policies?

2. Does the organization rely too much on rules and procedures?

3. Are the policies designed and communicated in a way that supports adherence?

4. Are the policies current and refreshed?

5. Are policies and relevant resources available and accessible?

6. Are there too many workarounds? If yes, what is the reason?

7. Is the allocation and distribution of workload appropriate? Consider work hours, shifts, rotations, rest periods, night work.

Recognize Emergent Factors

Workload

1. Is there an increased demand for work that is beyond the normal to any of the involved healthcare professionals?

2. Are the people involved working more than normal?

Distractions and Interruptions

1. Does the work environment introduce distractions or interruptions?

2. Are the distractions or interruptions frequent?

3. Do people seem to be affected by the interruptions or distractions?

Mental and Physical Workload

Mental Workload

1. Do the perceived cognitive demands of the task or the environment exceed one's cognitive capabilities?

2. Which cognitive capabilities are overloaded?

Physical Workload

1. Do the physical demands of the task or the environment exceed one's physical capabilities?

2. What specific physical actions are overloaded?

Discomfort

1. Can the available space result in physical discomfort, pain, or harm?

2. Can any other factor result in physical discomfort, pain, or harm?

Fatigue and Loss of Vigilance

1. Is one required to sustain a high workload or stressful situation for a prolonged period?

2. Are there symptoms of fatigue?

3. Is one required to sustain a very low workload for a prolonged period?

Stress

1. Are there environmental—physical and/or human/organizational—stressors?

2. Do these stressors push the individual or team beyond their capabilities?

3. What are the stress symptoms?

4. Do the people handle stress?

CONCLUDE: PERFORMANCE AND OUTCOMES

1. Can we explain what happened? Or, can we explain what may happen?

2. What should we do about it?

Outcomes: Factual, Likely, and Desired

Patient-specific Outcomes

- How did the performance of the unit of analysis affect the patient's health (clinical outcomes)?

- Was the patient satisfied with the performance?

- What were the monetary costs involved?

Healthcare Professional Outcomes

- What was the impact of performance on the healthcare professional?

- What emotional impact did they experience?

- Are healthcare professionals satisfied with their work?

Other People Outcomes

- Were the other people satisfied with the performance?

- What monetary costs were involved?

Organizational Outcomes

- What operational impacts did the performance have, such as wait times, discharge times, and hospitalization times?

- What monetary costs were involved?

- What legal consequences were there?

- How did performance influence the organization's reputation?

Appendix B: Fully Analyzed Sepsis Management Scenario

The Sample Scenario

Sepsis Management Scenario:

Heather, a 47-year-old woman, is quite healthy at baseline. She has had several episodes of renal colic (flank pain from kidney stones) throughout her life, and underwent laparoscopic removal of a large right-sided kidney stone 4 or 5 years ago.

Heather came to the Emergency Department at 6:30 AM after experiencing the too-familiar right-sided flank pain for several days. She entered the ED's main entrance and somehow managed to pull herself over to the registration cubicle. She was somewhat confused, and was not able to remember her home address or phone number. Her first recorded vital signs were as follows: HR, 128; BP, 100/65; Temp, 38.8 0C; RR, 24; O_2Sat, 94% on room air.

Heather ended up waiting for 45 min before she was registered. The triage nurse did not screen her for sepsis, and did not appreciate that she was very ill. She ended up staying in the waiting area for an additional 1.5 hr, receiving no care at all. She then collapsed—fell on the floor, lost consciousness—in the waiting area, and this prompted the staff to bring her urgently to one of the resuscitation rooms in "Section A." At this stage she required aggressive resuscitation, including intubation (of the windpipe), mechanical ventilation, urgent placement of central venous catheter, and admission to the ICU. It took 4 hr before the first dose of antibiotics was given, and blood cultures were not drawn at all. The delay in managing her resulted in acute kidney failure, and she required hemodialysis for several weeks. She ended up requiring critical care for 3 weeks, then 2 additional weeks on a regular hospital ward, and finally rehab at the local center.

Determine Objectives and Scope of the Analysis

Objectives: Analysis of an Adverse Event

We want to understand why the patient was not screened for sepsis, what were the causes for delays throughout the management of the case, and why did the case end up with outcomes that included resuscitation, ICU, hemodialysis, and prolonged critical care?

The Units of Analyses

The ED team at the time with a focus on the triage nurse

Stakeholders

ICU directors, hospital management

Map the Context

People

Here is an example of mapping people in the sepsis scenario.

| People | Characteristics | | Context | Role | Level of Involvement |
	Profession, Qualifications	Knowledge, Experience			
Triage RN	Nursing school (RN); Emergency Medicine training certificate	Full time; several years of experience in Emergency Medicine; variable experience in triage role	ED – Triage area	Triage	Primary
Unit Clerk	Medical Unit Clerk training	Full time; years of experience	ED – Triage area	Administrative	Secondary
Clinical Area RN	Nursing school (RN); Emergency Medicine training certificate	Full time; several years of experience in Emergency Medicine	ED – clinical area	Management in ED	Primary
Attending physician	MD; Emergency Medicine training program (5 years)	Full time; several years of experience; highly skilled in evaluation and resuscitation of acutely ill patients	ED – clinical area	Management in ED	Primary

Missions, Goals, Tasks

The following is Sepsis Management Scenario—Mapping Goals and Tasks.

People	Goals and Tasks Mapping		
	Goals	Tasks	Required capabilities
Triage RN	Identify sepsis; expedite management	Take history, take vitals, suspect sepsis, activate "Sepsis protocol" (i.e., use the sepsis workbook), including communicate with MD, facilitate initial lab tests, expedite patient transfer to clinical area, communicate with clinical area RN	Be familiar with the sepsis syndrome; be able to use the sepsis workbook; communication skills, in particular be able to communicate urgency, skilled in closed-loop communication techniques; multi-tasking; attention allocation/division; computer and other device literacy; work with other people; ability to adapt and change actions and directions; manage pressure and urgency, time pressures;
Clerk	Register the patient	Register patient in the electronic system; identify an unwell patient; communicate concern to triage nurse	Basic clinical skills, communication skills—assess urgency
Clinical Area RN	Adequately provide all elements of sepsis care	Establish IV access; draw blood for tests; administer fluids and medications; monitor patient; communicate concerns to MD	Technical skills (IV access, infusion, etc.); other clinical skills (identifying an unwell patient); communication skills (urgency, closed loop); multi-tasking; attention allocation/division; computer and other device literacy; leadership; work with other people; ability to adapt and change actions and directions; manage pressure and urgency, time pressures;

| MD | Identify/diagnose sepsis; adequately manage a septic patient | Take history and perform focused physical exam; order appropriate initial fluid resuscitation, lab and other tests, and antibiotics; monitor response to treatment and respond appropriately; insert a central venous catheter if needed | Be familiar with the sepsis syndrome; skilled in resuscitation, including technical skills (central line insertion, bedside U/S); communication skills—perceive urgency from RN; multi-tasking; attention allocation/division; computer and other devices literacy; leadership; work with other people; ability to adapt and change actions and directions; manage pressure and urgency, time pressures |

Task Analysis – Map the Inter-relations between the Tasks

The following task analysis is a normative example when sepsis may be involved, that is, it describes how sepsis should be done.

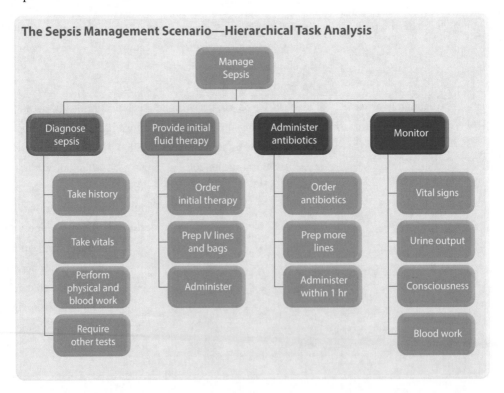

The Sepsis Management Scenario—Hierarchical Task Analysis

Task Flow Mapping

The following diagram is a swim lanes diagram for the sepsis management scenario.

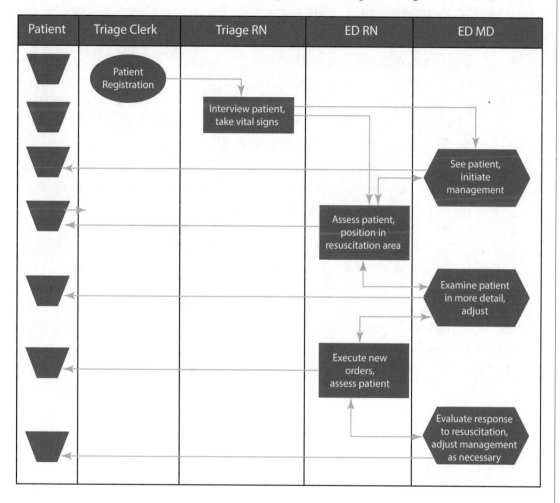

Physical Environment

Space and Layout

Here are answers for the sepsis management scenario. The link analysis diagram follows.

- The ED comprises a large waiting area and several clinical sections where care is provided. The registration and triage processes take place in the waiting area. A patient arriving independently to the ED would first register with the clerk, and then approach the adjacent Triage desk to be seen by a nurse. If no emergent condition is identified, the patient will likely be instructed to take a seat and wait to be called in. There are about 20–25 seats available in the waiting area, a hall of roughly 1,000 ft^2.

- A large door separates the triage from the clinical areas.

- The clinical management is provided in 4–5 sections (A–E), arranged in a circle around a common corridor, and each contains about 10–12 stretchers. "Section A," the "high acuity" area, is the closest to the main entrance and triage area. It contains three "Resuscitation Rooms," where equipment necessary for various life-sustaining therapies is stocked routinely. The Charge Nurse is usually present in Section A. Typically the nursing stations are located at the center of the different areas, surrounded by the patients' stretchers. There are multiple rooms and cabinets with medical equipment, supplies and medications, several washrooms, and working stations for clinicians. There is an X-Ray room at about the center of the ED, however, CT and other radiology services are located outside of the ED.

- The ED the waiting area was quite full of people, patients, and healthcare workers.

Using Link Analysis to Analyze the Physical Environment

Here is a spaghetti diagram based on the sepsis management scenario.

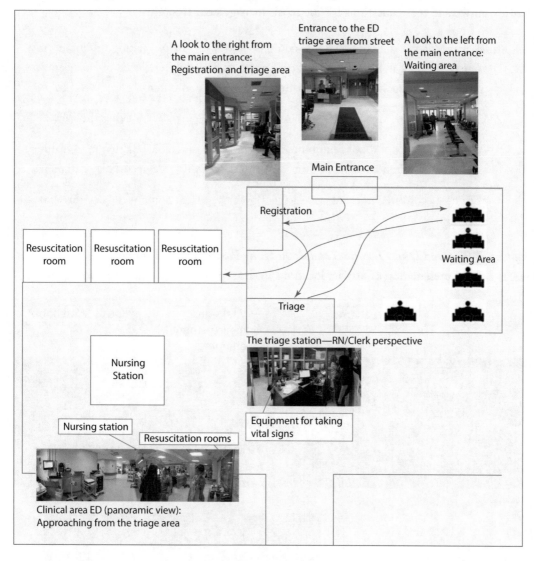

Artifact and Device Locations and Access

Mapping Artifact and Device Locations and Access Using the Detailed Checklist
Here are answers to these questions for the sepsis management scenario.

1. Triage desk for vital signs taking, computer, docs for history taking; phone; forms; same equipment in the ED itself if patient taken directly there; used by nurse and/or physician; in triage all on desk next to nurse; in ED, computer.

2. Housekeeping big cart was located just at the center, partially blocking the way.

3. Artifacts: sepsis workbook, equipment for vital signs measures (BP, temp, O_2 saturation), monitor, fluid bags, lines, drugs, O_2 supplementation devices, Foley catheters.

4. Various pieces of equipment were left all over the place. Some were still dirty after being used.

Mapping Artifact and Device Locations and Access Using Pictures and Diagrams
Here is a visual presentation of artifact locations in the sepsis management scenario.

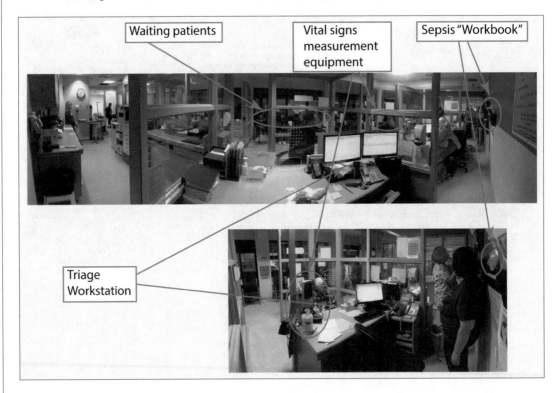

Ambient Conditions

The following are answers to the ambient condition questions for the sepsis management.

1. Fluorescent light is used throughout the triage area and the main clinical area; little daylight penetrates the triage and waiting area.

2. Noise can reach high levels. The main source of noise is the usual commotion such as people talking, phones ringing, pagers beeping, some medical devices such as monitors and infusion pumps, and other office machinery such as printers.

3. Temperature is controlled through air conditioning throughout the ED.

4. Air quality is good

5. No vibrations exist

6. Odors are not common, but may include human scents—natural and artificial.

Human-Machine Interface

The following are answers to the HMI checklist display questions for the sepsis management scenari.

1. Usual electronic medical record content.

2. Demographic and clinical information.

3. Depending on the information type, some is quite detailed, other may be general content.

4. After registration, information may be updated as frequently as new information such as vital signs, assessment and test results is generated.

5. Manually entering erroneous information (by human operators) is possible. Intercepting an error is up to the individual. Test results are mirrored in the system as reported by the lab.

6. Information may be entered manually by different operators or transmitted electronically from different users such as the lab and the imaging department.

7. At least two ways—manual entering and electronic transfer.

8. Information is presented on a computer screen.

9. N/A

10. To enter or review information about a patient, the provider has to log on to the system and to open up the specific patient's electronic chart.

Controls

1. Entering and/or accessing information.

2. Fingers and hand.

3. Keyboard and mouse.

Dialogue

1. By logging on to the system and to the specific patient's chart.

2. The dialogue is discreet. Users typically review and/or enter information repeatedly over the process of managing the patient.

3. Once information is entered the user cannot undo it. There is an option to add a comment regarding erroneous information.

4. Information can be accessed and observed as per the user's will. Users can enter information using mostly free text and sometimes selecting from a menu.

5. The dialogue is completed by saving an entry and logging off. Not uncommonly, a user would abort the dialogue through just walking away from the workstation.

Human Environment

Here is an example of aspects of groups and teams in the sepsis management scenario.

1. In the triage area: patients and care providers, registration staff, triage nurses, security staff, volunteers.

2. Patients and escort: waiting to be seen; talking—over the phone or among themselves; staff members: attend to their respective roles in the area, including charting, interviewing, giving instructions, etc.

3. Stakeholders compose of individuals with ad-hoc teams. Typically, a single registration clerk or two; one or two triage nurses; multiple ED nurses and physicians, including attending and resident physicians at different levels.

4. No real hierarchy among nurses and registration clerks; no hierarchy among triage and clinical area nurses; doctors in a position to give orders, make decisions, and thus perceived hierarchical gradient between doctors and nurses.

5. Triage nurses, out of the nature of their activities (e.g., assigning illness severity to patients, prioritizing care) are the natural leaders in the triage area. Yet, leadership role may not be formalized and accepted by all.

6. Providers usually know each other, at least within profession (i.e., clerks, nurses, doctors). There may be newer providers that may not know others/vice versa. Obviously, providers usually do not know the patients prior to their admission.

7. Communication is usually ad hoc, and verbal, face to face. Phone is also an option.

8. Coordination and collaboration usually occurs verbally, on the fly.

9. Roles are clearly defined as per the profession (e.g., clerks, nurses, doctors)

10. Workload varies and is for the most part unpredictable. Some hours in the day may be quiet and others very busy. Similarly, some particular events such as acute stroke protocol and trauma team activation may suddenly and substantially increase workload.

11. For the most part, providers have a good understanding of their own roles and their colleagues' roles. In the case of sepsis protocol, once the protocol is initiated, providers usually have a good understanding of their roles.

Organizations and Organizational Culture

Here is an example of aspects of organizations and organizational culture in the sepsis management scenario.

1. There are role definitions and responsibilities.

2. Nurses and clerks are hospital employees. Doctors are not; however, they are granted privileges to operate within the hospital, and they are to follow hospital policies.

3. Sepsis protocol is often not followed. In the triage area there will be one or two clerks at a given time, and one or two nurses. There are no doctors in the triage area. The ED's main clinical area is staffed by clerks, nurses, doctors, and other staff members.

4. Employees are expected to follow hospital policies; they are held accountable for their actions.

5. Learning takes place ad hoc, especially when a new device or process is introduced. Time constraints and other considerations preclude effective learning quite often, and thus staff would often learn on the fly.

6. Email is commonly used. In-person interaction is less common.

7. Development and further training is for the most part to the discretion of the employee (in the case of nurses and clerks).

Rules, Regulations, and Policies

Here is an example of aspects of rules, regulations, and policies in the sepsis management scenario.

1. The sepsis protocol is available as an electronic file within the hospital's intranet, and also in a physical binder in different locations in the ED.

2. The sepsis protocol was designed by a multi-disciplinary team, including ED providers.

3. Teaching sessions were held for most doctors and about two thirds of the nurses. It was almost communicated often over email. A survey of ED providers revealed that the protocol is well received and easy to use. However, often users forget to use it, and they can't find it.

4. Nurses work 12-hr shifts. The triage position is often rotated throughout the shift. Breaks are scheduled into the work day. Doctors work 8-hr shifts.

5. Providers are generally disciplined and respect rules and policies. However, the prevailing rules may represent culture more than written policy, which providers usually are unaware of.

Interim Brief #1

Here is the key mapping elements table, filled in with the sepsis management scenario.

Unit of Analysis	Factors	Sub-factors	Key Mapped Factors
	People		Physicians, nurses, clerks
	Goals and Tasks		Goals: Providing adequate management for septic patients. Main tasks: Diagnose sepsis, provide initial fluid therapy, administer antibiotics, monitor
	Physical Environment	Physical space	Triage area followed by direct access and passage to the ED's main clinical area
		Layout of sub-spaces and general locations	Busy waiting and triage area; separate from the main clinical area; space for patients to sit down (disappear?)
		Artifact and device locations and access	Face-to-face interaction with patient; vital sign measurement equipment; sepsis workbook (protocol) in a binder at the side and behind
		Ambient conditions	Noise; fluorescent lights; air conditioned
		HMI of devices and tools	Usual computer workstations; electronic health record, info entered by different people at different places
	Human Environment	Teams and groups	Skilled providers; roles defined; yet, team structure, including leadership roles, not defined
		Organizations and organizational culture	Sepsis protocol often not followed; difficulty allocating time/resources for education; difficulty maintaining continuous learning
		Rules, regulations, and policies	Written policy usually not followed. Strong culture of good patient care

Assess Fit

People and Tasks

The People Perspective

The following answers assess the fit of the relevant people to tasks in the sepsis management scenario.

Checklist Item	Data	Fit Score
1	The nurses may not have a good knowledge and appreciation of the management of septic patients, particularly in recognizing a sepsis patient.	4
2	Motivation is high.	7
3	In general, the nurses are committed to providing high quality care.	6
4	The nurses are aware of some issues in their work environment. However, they may get used to some hazards over time.	4
5	The nurses usually don't take unnecessary risks knowingly. However, they may unknowingly get used to unsafe practices attempting to improve efficiency, or through other mechanisms.	5
6	For the most part, the nurses feel they are in control. A busy environment may change that.	5
7	In general, the nurses handle stress well.	6
8	In general, the ED nurses are nervous.	6
9	Usually the nurses can divide their attention.	6
10	The nurses make decisions mostly intuitively, or in a rule-based fashion, and usually effectively.	5
11	In the context of sepsis, cognitive biases may be common. Confirmation and anchoring biases may contribute to misdiagnosis of sepsis, and thus to failure to manage it properly.	3

The Task Perspective

The following answers assess the fit of the tasks to the relevant unit(s) of analysis (primarily nurses) in the sepsis management scenario.

Checklist Item	Data	Fit Score
1	The number of tasks is manageable.	7
2	Especially at times of peak activity, nurses may need to address too many tasks concurrently. Usually, however, this is not a problem.	6
3	There is sufficient time to perform the task, once the need is recognized.	7
4	The tasks are quite straightforward and easy to execute.	7
5	The necessary information is quite basic and easy to obtain.	7
6	For the most part, dividing attention to the tasks at hand is quite easy.	6
7	Focusing attention on the relevant information could be challenging at times.	4
8	Technically, it is easy to track and monitor information over time. The problem is more on the side of remembering to do so.	6
9	Technically, the time available is sufficient for making adequate decisions.	6
10	N/A	
11	Physical effort is appropriate.	7
12	Physical effort does not challenge the respiratory or cardio-vascular system.	7

Physical Environment

Space and Layout

The following answers assess the fit of the space and layout to the people (the relevant unit(s) of analysis—primarily nurses) and the tasks in the sepsis management scenario.

Checklist Item	Data	Fit Score
1	With recently increasing crowding in the ED, it is not uncommon that space is limited.	5
2	For the most part space is sufficient to perform tasks.	6
3	Space configuration can pose challenges to the workflow. It may be difficult to see a patient; patients need to be transferred from the triage area to the main ED space to be treated.	4
4	Quite often access paths are crowded.	4
5	Entry and exit point locations are less relevant to the sepsis workflow.	7

Location of Artifacts and Devices

The following answers assess the fit of the space and layout to the people (the relevant unit(s) of analysis—primarily nurses) and the tasks in the sepsis management scenario.

Checklist Item	Data	Fit Score
1	Vital sign measurement devices, as well as the sepsis workbook, are located at the triage booths.	6
2	Sepsis workbook is not at point of care; However, location of other artifacts less relevant to this scenario, as it is uncommon for the triage nurses to encounter multiple septic patients concurrently.	6
3	It is very easy to lose track of a patient, and his/her information.	2
4	Lack of focused attention in the visual search could result in failure to identify important information.	2
5	It is not very difficult to divide attention between the sources of information. The issue is focusing attention on the right information.	5
6	Information per se is displayed legibly. However, the nurse needs to initiate seeking information.	5
7	Nurses often need to get up from their chair and move around the space to get necessary information and artifacts.	2

Ambient Conditions

The following answers assess the fit of the ambient conditions to the people (the relevant unit(s) of analysis—primarily nurses) and the tasks in the sepsis management scenario.

Checklist Item	Data	Fit Score
1	Light is adequate; there are some reflections from glass windows and partitions.	6
2	Lighting conditions not an issue.	7
3	The triage area can be noisy; distractions and interruptions are common.	4
4	For the most part, noise distractions and interruptions are not a problem.	6
5	While distinguishing between different sounds may be challenging in the triage area, this is less of an issue for the sepsis management task.	6
6	N/A	
7	N/A	
8	N/A	
9	The area is air conditioned and comfortable.	7
10	N/A	
11	N/A	
12	Intense odors are uncommon.	6

Device Usability

The following table evaluates the device usability of the electronic health records (EHR) for the sepsis management scenario.

Statement	Strongly Disagree			Undecided			Strongly Agree
	(1)	(2)	(3)	(4)	(5)	(6)	(7)
The work flow has a clear beginning and end. them; not pushed. Delays in flow of info from the room into the EDIS							
All functions are grouped to support the work flow. User has to interact with several different systems The systems are very different; emergency has another system for X-Ray results				✓			
It keeps the user informed about what goes on.				✓			
The information appears in natural and logical order.				✓			
The device allows the user to have full control.				✓			
The device prevents errors from occurring. There could be omission errors				✓			
The device provides clear ways to undo actions and recover from errors.				✓			
The device "speaks" the users' language. EDIS gives the blood test results without trends, only numbers; trend is very important, and for that one needs to log into another system, PCS				✓			
The device looks aesthetically pleasing.				✓			
The device has a simple visual design. Color coding of monitor may not be clear to everyone				✓			

Human Environment

Groups and Teams

The following answers assess the fit of the space and layout to the people (the relevant unit(s) of analysis—primarily nurses) and the tasks in the sepsis management scenario.

Checklist Item	Data	Fit Score
1	Presence of people is usually not an issue. Occasionally, in a very busy environment, workflow may slow down.	6
2	During usual workload nurses can divide attention well between relevant people and tasks.	6
3	The team at the triage area is quite small, consisting of 1-2 nurses and 1-2 clerks.	6
4	It is unclear whether a leader has been assigned in the triage area.	4
5	Nurses attend only to nursing tasks; clerks attend only to their own tasks.	1
6	It is hard to tell whether communication is effective. There are no clear expectations for communication, and decisions are left in the hands of members.	4
7	It is not clear if they have shared mental models.	4
8	It is not clear if they have shared situational awareness.	4
9	The challenge is more in failing to communicate.	7
10	Once an issue is communicated, there is pretty good coordination and collaboration among members.	6
11	There is a clear definition of roles.	7
12	The allocation of workload can be a hit or miss. When the environment is busy, typically everyone would work harder.	6
13	People have good structured knowledge of who does what, when, and how.	7
14	Team members share similar expectations and understanding of everyone's role and what is supposed to be done when and how.	7
15	Team members are aware of what is going on and sometimes communicate it.	6
16	It is easy to lose situational awareness and not share it, especially in a busy environment.	4

Organizations, Climates, and Cultures

The following answers assess the fit of the organization, climate, and culture to the people (the relevant unit(s) of analysis—primarily nurses) and the tasks in the sepsis management scenario.

Checklist Item	Data	Fit Score
1	Division of roles within the organization is clear.	7
2	The organization structure supports achieving the goals.	7
3	Staffing is appropriate.	7
4	Employee surveys have revealed some degree of dissatisfaction among staff regarding support from leadership.	4
5	Members of the organization are involved and engaged for the most part.	6
6	Safety culture surveys have repeatedly indicated suboptimal perceptions of staff regarding safety and quality.	4
7	The term 'accountability' often has a negative connotation, implying 'you better do your job!'	3
8	Learning toward improved practice is uncommon.	2
9	Communication is based on intranet/email, which fail to reach many.	2
10	Generally, team members are adept in resuscitation. The missing training aspect is in recognizing a septic patient.	6

Rules, Regulations, and Policies

The following answers assess the fit of the organization, climate, and culture to the people (the relevant unit(s) of analysis—primarily nurses) and the tasks in the sepsis management scenario.

Checklist Item	Data	Fit Score
1	There are sufficient and appropriate policies.	7
2	There are very limited activities to ensure that policies, rules, and procedures are actually followed.	1
3	Policies are not designed and communicated in a way that supports adherence.	1
4	Policies are current for the most part.	6
5	Policy resources are available; however, staff usually doesn't seek to find and read them.	4
6	It is unknown if there are too many workarounds. In the case of sepsis, workarounds are likely not a substantial issue.	5
7	For the most part workload is appropriate.	6

Interim Findings: Problems of Fit

List of Fit Problems by Factors

The following table is an example of the summary of fit assessment for the sepsis management scenario.

Unit of Analysis	Context		Fit with capabilities and limitations	Fit Score
	Factors	**Sub-factors**		
Triage area, including a unit clerk and a triage nurse	Goals and Tasks		The task of identifying and initiating the workflow for septic patients has a good fit with the capabilities of involved team members. However, it is quite likely that nurses often lack the expertise to recognize a septic patient, especially in early stages of illness. Focusing on the right information is challenging. That said, providers are highly motivated to do a good job. Cognitive biases may occur.	4
	Physical Environment	Physical space	The area is crowded.	4
		Layout of sub-spaces and general locations	The physical space and crowdedness pose challenges to providers' ability to observe and monitor a septic patient.	3
		Artifact and device locations and access	The sepsis workbook (protocol) is unavailable at the point of care; nurses often need to make extra effort to access relevant information.	3
		Ambient conditions	Although crowding and noise can be a nuisance at the triage area, the ambient conditions are comfortable for the most part.	5
		HMI of devices and tools	The logical grouping and consistency of information relevant to sepsis in the EHR is inappropriate.	4

Unit of Analysis	Context		Fit with capabilities and limitations	Fit Score
	Factors	Sub-factors		
Triage area, including a unit clerk and a triage nurse	Human environment	Teams and groups	There is no clearly defined leader. It is not clear if communication is always effective, if there is shared understanding at all times, and if situational awareness is degraded	3
		Organizations and organizational culture	There is some dissatisfaction with leadership. The safety climate is insufficient. An "accountability" culture is perceived negatively. There is insufficient learning to improve. Within an organization communication does not reach everyone.	3
		Rules, regulations, and policies	Mechanisms to ensure adherence to policies and procedures are insufficient. Policies are not implemented appropriately. Staff do not always seek the relevant policies.	4

Problem Severity

The following table is an example of the summary of severity of problems in the sepsis management scenario.

	Problem area	Problems	Severity
Goals and Tasks	Goals		
	Tasks	Variable level of skill among nurses in recognizing that a patient is septic: Fairly common; when present likely to be a problem with major impact on outcomes	High
Physical Environment	Physical space	Crowdedness	Medium
	Layout of sub-spaces and general locations	Challenges to providers' ability to observe and monitor a septic patient	Medium
	Artifact and device locations and access	Sepsis workbook not present: Uncommon; when missing, can be accessed somewhere in the ED; good clinical skills can negate the need for it.	Low

	Problem area	Problems	Severity
Physical Environment		Lack of readily available information, and alerts regarding, abnormal physical and lab-based findings: Common, substantial impact on outcomes.	High
	Ambient conditions	Noise	Low
	Usability of devices and tools	Inconsistent and hard to reach information	Medium
Human Environment	Teams and groups	Once sepsis is recognized, team members communicate well regarding sick patients. However, failing to close communication loops may affect, likely commonly, and with substantial consequences	High
	Organizations and organizational culture	Poor capability to promote learning in the organization can have major consequences	High
	Rules, regulations, and policies	Problem with adopting and adhering to policies and regulations	Low

Recognize Emergent Factors

Emergent Environmental Factors

Workload

The following answers assess workload of the people (the relevant unit(s) of analysis—primarily nurses) and the tasks in the sepsis management scenario.

1. Sepsis management does not typically increase workload significantly.

2. The healthcare professionals involved in sepsis management do not work significantly more than normal.

Distractions and Interruptions

The following answers assess distractions and interruptions of the people (the relevant unit(s) of analysis—primarily nurses) and the tasks in the sepsis management scenario.

1. The work environment does introduce distractions and interruptions.

2. The distractions and interruptions could be high at times.

3. Using the sepsis protocol can degrade the impact of the distraction and interruptions. If not, there could be adverse impact of the distractions and interruptions.

Emergent Human Factors

Mental and Physical Workload

Mental Workload

The following answers assess the mental workload of the people (the relevant unit(s) of analysis—primarily nurses) in the sepsis management scenario.

1. Cognitive capabilities are not overloaded, particularly when using the sepsis protocol. Not using the protocol may add substantially to providers' cognitive load in having to remember the different management elements, and following up on the different indicators of a patient's response to treatment. However, the sometimes lack of relevant skills in recognizing sepsis may increase mental workload, particularly in the subsequent procedure once it is recognized. In addition, ineffective communication may also increase mental workload.

2. Some challenges to cognitive capabilities can rise due to the unclear and inconsistent presentation of relevant information in the EHR.

Physical Workload

The following answers assess the physical workload of the people (the relevant unit(s) of analysis—primarily nurses) in the sepsis management scenario.

1. Typically, physical demands of managing sepsis do not exceed physical capabilities.

2. Due to the layout, some physical demands may increase when trying to follow and monitor the patient.

Discomfort

The following answers assess the discomfort of the people (the relevant unit(s) of analysis—primarily nurses) in the sepsis management scenario.

1. No discomfort results from being engaged in sepsis management.

2. The difficulty in accessing relevant information or the sepsis workbook can result in some frustration.

Fatigue and Loss of Vigilance

The following answers illustrate that, sometimes, the questions may not be applicable, as in the case for assessing the fatigue and loss of vigilance of the people (the relevant unit(s) of analysis—primarily nurses) in the sepsis management scenario.

1. N/A

2. N/A

3. N/A

Stress

The following answers assess the fatigue and loss of vigilance of the people (the relevant unit(s) of analysis—primarily nurses) in the sepsis management scenario.

1. Since sepsis management, within the scenario, takes place in the ED, there are many stressors typical to the ED environment such as noise, clutter, over-crowdedness, too many patients, too many tasks.

2. The individual or team engaged in sepsis management can be pushed beyond their capabilities, given the presence of the stressors common to the ED environment.

3. In this case, the stress signs included a lack of patience, one nurse left the premises for 10 min. Another nurse cried.

4. The nurses' ability to handle the situation varied among team members. The triage nurse was quite discouraged with the queue of patients he still had to interview and examine.

Summary of Emergent Factors

The following table is an example of the summary of severity of emergent factors in the sepsis management scenario.

Key Factors	Problem area	Problems	Severity	Emergent Factors	Severity
Goals and Tasks	Goals				
	Tasks	Variable level of skill among nurses in recognizing that a patient is septic: Fairly common; when present likely to be a problem with major impact on outcomes	High	Increased mental workload	High
Physical Environment	Physical space	Crowdedness	Medium	Some stress	Medium
	Layout of sub-spaces and general locations	Challenges to providers' ability to observe and monitor a septic patient	Medium	Some discomfort and frustration; some stress	Medium
	Artifact and device locations and access	Sepsis workbook not present: Uncommon; when missing, can be accessed somewhere in the ED; good clinical skills can negate the need for it.	Low	Increased mental workload; some discomfort and frustration; little stress	Medium
		Lack of readily available information and alerts regarding abnormal physical and lab-based findings: Common, substantial impact on outcomes	High	High stress	Medium
	Ambient conditions	Noise	Low	Increase mental workload; little stress	High
	Usability of devices and tools	Inconsistent and hard to reach information	Medium	Increase mental workload; some stress	Medium

Key Factors	Problem area	Problems	Severity	Emergent Factors	Severity
Human Environment	Teams and groups	Once sepsis is recognized, team members communicate well regarding sick patients. However, failing to close communication loops may affect, likely commonly, and with substantial consequences	High	Increase in dissatisfaction and frustration	Medium
	Organizations and organizational culture	Poor capability to promote learning in the organization can have major consequences	High	Increase mental workload; high stress	Medium
	Rules, regulations, and policies	Problem with adopting and adhering to policies and regulations	Low	Little stress	Medium

Conclude: Performance and Outcomes

Identify the Most Influential Factors

Here are the influential factors determined in the sepsis management scenario. Note that factors that did not get identical severity ratings for the problems and emergent factors, the relevant factor was nevertheless considered an influential factor even if only one of the ratings was high.

Key Factors	Problem area	Problems	Severity	Emergent Factors	Severity	Influential Factors
Goals and Tasks	Goals					
	Tasks	Variable level of skill among nurses in recognizing that a patient is septic: Fairly common; when present likely to be a problem with major impact on outcomes	High	Increased mental workload	High	Insufficient experience in sepsis management

Key Factors	Problem area	Problems	Severity	Emergent Factors	Severity	Influential Factors
Physical Environment	Physical space	Crowdedness	Medium	Some stress	Medium	Crowdedness
	Layout of sub-spaces and general locations	Challenges to providers' ability to observe and monitor a septic patient	Medium	Some discomfort and frustration; some stress	Medium	Layout—insufficient support of task
	Artifact and device locations and access	Sepsis workbook not present: Uncommon; when missing, can be accessed somewhere in the ED; good clinical skills can negate the need for it.	Low	Increased mental workload; some discomfort and frustration; little stress	Medium	Inappropriate location of sepsis workbook;
		Lack of readily available information and alerts regarding abnormal physical and lab-based findings: Common, substantial impact on outcomes	High	High stress	Medium	Inappropriate location of other information
	Ambient conditions	Noise	Low	Increase mental workload; little stress	High	Noise
	Usability of devices and tools	Inconsistent and hard to reach information	Medium	Increase mental workload; some stress	Medium	Poor usability of EHR in support of the task

Key Factors	Problem area	Problems	Severity	Emergent Factors	Severity	
Human Environment	Teams and groups	Once sepsis is recognized, team members communicate well regarding sick patients. However, failing to close communication loops may affect, likely commonly, and with substantial consequences	High	Increase in dissatisfaction and frustration	Medium	Ineffective communication
	Organizations and organizational culture	Poor capability to promote learning in the organization can have major consequences	High	Increase mental workload; high stress	Medium	Insufficient learning and professional development
	Rules, regulations, and policies	Problem with adopting and adhering to policies and regulations	Low	Little stress	Medium	Ineffective policy implementation

Performance and Outcomes – How are They Different?

Key Factors	Problem area	Influential Factors	Performance
Goals and Tasks	Goals	Insufficient experience in sepsis management	Likely errors degrading effectiveness and causing delays
	Tasks		
Physical Environment	Physical space	Crowdedness	
	Layout of sub-spaces and general locations	Layout insufficient support of task	Delays
	Artifact and device locations and access	Inappropriate location of sepsis workbook	
	Ambient conditions	Noise	Degrading effectiveness
	Usability of devices and tools	Poor usability of EHR in support of the task	Degrading effectiveness and causing delays

Key Factors	Problem area	Influential Factors	Performance
Human Environment	Teams and groups	Ineffective communication	Likely errors degrading effectiveness and causing delays
	Organizations and organizational culture	Insufficient learning and professional development	Degrading effectiveness
	Rules, regulations, and policies	Ineffective policy implementation	Degrading effectiveness

Outcomes: Factual, Likely, and Desired

The following table is the summary of outcomes for the sepsis management scenario.

Key Factors	Problem area	Influential Factors	Performance	Outcomes
Goals and Tasks	Goals			
	Tasks	Crowdedness	Likely errors degrading effectiveness and causing delays	Longer recovery time and hospital admission
Physical Environment	Physical space	Crowdedness		Increased morbidity and mortality
	Layout of sub-spaces and general locations	Layout insufficient support of task	Delays	Poor patient and caregiver experience and satisfaction, loss of trust in the care team, frustration, and financial losses
	Artifact and device locations and access	Inappropriate location of sepsis workbook; Inappropriate location of other information	Likely errors degrading effectiveness and causing delays	Increased provider workload and stress, potential guilt, moral distress, and burnout
	Ambient conditions	Noise	Degrading effectiveness	Increased wait times in the ED, preventable ICU admissions, and longer ICU and hospital stays
	Usability of devices and tools	Poor usability of EHR in support of the task	Degrading effectiveness and causing delays	

Key Factors	Problem area	Influential Factors	Performance	Outcomes
Human Environment	Teams and groups	Ineffective communication	Likely errors degrading effectiveness and causing delays	Increased cost to the hospital and health-care system
	Organizations and organizational culture	Insufficient learning and professional development	Degrading effectiveness	Risk of litigation and damaged reputation for the hospital
	Rules, regulations, and policies	Ineffective policy implementation	Degrading effectiveness	

Interventions and Mitigations

Granularity of the Recommendations

Below is an example of the three granularity levels in Figure 11.1.

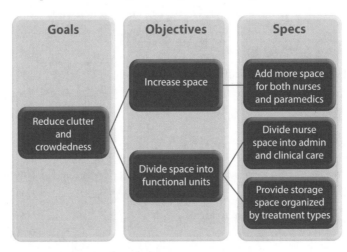

Intervention and Mitigation Strategic Goals

The following table provides sample interventions and mitigations for each problem area and its influential factors in the sepsis management scenario.

Key Factors	Problem area	Influential Factors	Interventions and Mitigations
Goals and Tasks	Goals		
	Tasks	Insufficient experience in sepsis management	Train staff (e-modules, small group learning, simulation).
Physical Environment	Physical space	Crowdedness	Keep working on strategies to alleviate bottlenecks and streamline patient flow through ED.
	Layout of sub-spaces and general locations	Layout insufficient support of task	As a new ED is being currently planned, engage in designing spaces that would support interaction and communication.
	Artifact and device locations and access	Inappropriate location of sepsis workbook; inappropriate location of other information	Workbook: Design a process to ensure that the workbook is available at all times. Other information: Difficult to redesign the current electronic chart. Potentially: work with the provider on introducing alerts (with a focus on usability) to providers while managing septic patients.
	Ambient conditions	Noise	Engage security as possible. Use technology to reduce noise.
	Usability of devices and tools	Poor usability of EHR in support of the task	See Other Information, above

Key Factors	Problem area	Influential Factors	Interventions and Mitigations
Human Environment	Teams and groups	Ineffective communication	Train staff in effective communication (issues: funding, time).
	Organizations and organizational culture	Insufficient learning and professional development	Same as above: the organization will need to find ways, and identify required resources, to ensure learning.
	Rules, regulations, and policies	Ineffective policy implementation	Same as above: the organization will need to identify strategies to facilitate policy implementation.

Prioritize the Interventions Using a Tradeoff Analysis

The following table is an example of the tradeoff analysis for the sepsis management scenario.

		Decision Criteria				
		Mitigation Feasibility	Mitigation Cost [5]	Mitigation Time[6]	Mitigation Success likelihood	
	Rank	3	2	2	4	11
	Relative	.27	.18	.18	.36	
	Insufficient experience	4 1.08	5 (1) .18	5 (1) .18	5 1.8	3.24
	Inappropriate location of workbook	5 1.35	2 (4) .72	2 (4) .72	5 1.8	4.59
	Ineffective communication	5 1.35	4 (2) .36	4 (2) .36	5 1.8	3.87
	Insufficient learning and development	3 .81	5 (1) .18	5 (1) .18	4 1.44	2.61

[5] Reverse scale
[6] Reverse scale

References and Resources

GENERAL HUMAN FACTORS

Boff, K. R. and Lincoln, J. E. (1988). *Engineering Data Compendium. Human Perception and Performance*. Harry G Armstrong Aerospace Medical Research Lab, Wright-Patterson AFB, Dayton, OH.

Carayon, P., Ed. (2011). *Handbook of Human Factors and Ergonomics in Health Care and Patient Safety*. CRC Press, Boca Raton, FL.

Gosbee, J. (2002). Human factors engineering and patient safety. *Qual Safe Health Care* 11.4, 352–354. DOI: 10.1136/qhc.11.4.352.

Porto, G. G. (2001). Safety by design: Ten lessons from human factors research. *J Healthc Risk Manag* 21.3, 43-50. DOI: 10.1002/jhrm.5600210408.

Sanders, M. S. and McCormick, E. J. (1993). Human Factors in Engineer and Design. McGraw-Hill.

Stanton, N. A., Young, M. S., and Harvey, C. (2014) *Guide to Methodology in Ergonomics: Designing for Human Use*. CRC Press, Boca Raton, FL. DOI: 10.1201/b17061.

Wickens, C. D., Lee, J., Liu, Y. D., and Gordon-Becker, S. (2013). *Introduction to Human Factors Engineering: Pearson New International Edition*. Pearson Higher Ed. 75

Wickens, C. D., Hollands, J. G., Banbury, S., and Parasuraman, R. (2015). *Engineering Psychology and Human Performance*. Psychology Press. 75

PATIENT SAFETY

Agency for Healthcare Research and Quality (AHRQ) (2001). *Making Health Care Safer: A Critical Analysis of Patient Safety Practices*. AHRQ, Rockville, MD.

Agency for Healthcare Research and Quality (2013). Module 4. Approaches to Quality Improvement. Content last reviewed May 2013. Agency for Healthcare Research and Quality, Rockville, MD. http://www.ahrq.gov/professionals/prevention-chronic-care/improve/system/pfhandbook/mod4.html. 4, 141

Ash, J. S., Berg, M., and Coiera, E. (2004a). Some unintended consequences of information technology in health care: the nature of patient care information system-related errors. *J Amer Med Inform Assn*, 11(2), 104–112. DOI: 10.1197/jamia.M1471. 4

Baker, G. R., Norton, P. G., Flintoft, V., Blais, R., Brown, A., Cox, J., and O'Beirne, M. (2004). The Canadian Adverse Events Study: the incidence of adverse events among hospital patients, in *Can Med Assn J*, 170(11), 1678–1686. DOI: 10.1503/cmaj.1040498. 3

Baker, G. R. and Black, G. (2015). *Beyond the Quick Fix: Strategies for Improving Patient Safety*. Institute of Health Policy, Management and Evaluation, University of Toronto. 3

Bogner, M. S. E. (1994). *Human Error in Medicine*. Lawrence Erlbaum Associates, Inc. 4

Carayon, P., Ed. (2016). *Handbook of Human Factors and Ergonomics in Health Care and Patient Safety*. CRC Press, Boca Raton, FL. 4

Carayon, P., Wetterneck, T. B., Rivera-Rodriguez, A. J., Hundt, A. S., Hoonakker, P., Holden, R., and Gurses, A. P. (2014a). Human factors systems approach to healthcare quality and patient safety. *Appl Ergono*, 45(1), 14–25. DOI: 10.1016/j.apergo.2013.04.023. 4

Christian, C. K., Gustafson, M. L., Roth, E. M., Sheridan, T. B., Gandhi, T. K., Dwyer, K., Zinner, M. J., and Dierks, M. M. (2006). A prospective study of patient safety in the operating room. *Surgery* 139.2, 159–173. DOI: 10.1016/j.surg.2005.07.037.

Cook, R. I. and Woods, D. D. (1994). Operating at the sharp end: the complexity of human error. *Human Error Med* 13, 225–310.

Cook, R. I., Render, M., and Woods, D. D. (2000). Gaps in the continuity of care and progress on patient safety. *Brit Med J*, 320(7237), 791. DOI: 10.1136/bmj.320.7237.791.

Cooper, J. B., Newbower, R. S., Long, C. D., and McPeek, B. (1978) Preventable anesthesia mishaps: a study of human factors. *Anesthesiology* 49.6, 399–406. DOI: 10.1097/00000542-197812000-00004.

Donchin, Y. and Seagull, F. J. (2002). The hostile environment of the intensive care unit. *Curr Opin Crit Care*, 8(4), 316–320. DOI: 10.1097/00075198-200208000-00008.

Gosbee, J. (2002). Human factors engineering and patient safety. *Qual Safe Health Care*, 11(4), 352–354. DOI: 10.1136/qhc.11.4.352. 4

Gurses, A. P., Ozok, A. A., and Pronovost, P. J. (2011). Time to accelerate integration of human factors and ergonomics in patient safety. *BMJ Qual and Safe*, bmjqs-2011. DOI: 10.1136/bmjqs-2011-000421.

Gurses, A. P. and Pronovost, P. J. (2011). Physical environment design for improving patient safety. *Health Environ Res Des J*, 5, 3–5. 4

Hughes, R., Ed. (2008). *Patient Safety and Quality: An Evidence-based Handbook for Nurses* (Vol. 3). Agency for Healthcare Research and Quality, Rockville MD. 4

Kohn, L. T., Corrigan, J. M., and Donaldson, M. S., Eds.). (2000). *To Err Is Human: Building a Safer Health System* (Vol. 6). National Academies Press. 3, 4

Langley, G. L., Moen, R., Nolan, K. M., Nolan, T. W., Norman, C. L., and Provost, L. P. (1996). *The Improvement Guide: A Practical Approach to Enhancing Organizational Performance.* Jossey-Bass , San Francisco. 5, 141

Leape, L. L. and Berwick, D. M. (2005). Five years after To Err Is Human: what have we learned?. *JAMA*, 293(19), 2384–2390. DOI: 10.1001/jama.293.19.2384.

Leape, L. L., Brennan, T. A., Laird, N., Lawthers, A. G., Localio, A. R., Barnes, B. A., Hebert, L., Newhouse, J. P., Weiler, P. C., and Hiatt, H. (1991). The nature of adverse events in hospitalized patients: results of the Harvard Medical Practice Study II. *NEJ Med*, 324(6), 377–384. DOI: 10.1056/NEJM199102073240605.

Lipsitz, L. A. (2012). Understanding health care as a complex system: the foundation for unintended consequences. *JAMA*, 308(3), 243–244. DOI: 10.1001/jama.2012.7551. 4

Makary, M. A. and Daniel, M. (2016). Medical error—the third leading cause of death in the US. *BMJ* , 353, i2139. DOI: 10.1136/bmj.i2139. 4

McDaniel, R. R., Driebe, D., and Lanham, H. J. (2013). Health care organizations as complex systems: new perspectives on design and management. *Adv Health Care Manag*, 15, 3–26. DOI: 10.1108/S1474-8231(2013)0000015007. 4

Pronovost, P., Weast, B., Rosenstein, B., Sexton, J. B., Holzmueller, C. G., Paine, L., Davis, R. and Rubin, H. R. (2005). Implementing and validating a comprehensive unit-based safety program. *J Patient Safety*, 1.1, 33–40. DOI: 10.1097/01209203-200503000-00008.

Pronovost, Peter J., Berenholtz, S. M., Goeschel, C., Thom, I., Watson, S. R., Holzmueller, C. G., Lyon, J. S., Lubomski, L. H., Thompson, D. A., Needham, D., Hyzy, R., Welsh, R., Roth, G., Bander, J., Morlock, L., and Sexton, J. B. (2008) Improving patient safety in intensive care units in Michigan. *J Crit Care*, 23.2, 207–221. DOI: 10.1016/j.jcrc.2007.09.002.

Sari, A. B., Sheldon, T. A., Cracknell, A., and Turnbull, A. (2007). Sensitivity of routine system for reporting patient safety incidents in an NHS hospital: retrospective patient case note review. *BMJ*, 334.7584, 79. DOI: 10.1136/bmj.39031.507153.AE.

Ting, H. H., Shojania, K. G., Montori, V. M., and Bradley, E. H. (2009). Quality improvement science and action. *Circulation*, 119(14), 1962–1974. DOI: 10.1161/CIRCULATIONAHA.108.768895. 4

Varkey, P., Reller, M. K., and Resar, R. K. (2007). Basics of quality improvement in health care. *Mayo Clinic Proc*, 82(6) 735–739). DOI: 10.1016/S0025-6196(11)61194-4. 4

Wachter, R. M. (2012). *Understanding Patient Safety*. McGraw Hill Medical, New York. 4

Weinger, M. B. and Slagle, J. (2002). Human factors research in anesthesia patient safety. *J Amer Med Info Assn*, 9 (Supplement 6), S58–S63. DOI: 10.1197/jamia.M1229. 4

Woods, D., Cook, R., Zipperer, L., and Cushman, S. (2001). From counting failures to anticipating risks: possible futures for patient safety. In *Lessons in Patient Safety*. National Patient Safety Foundation, Chicago, pp. 89–97. 4

HUMAN FACTORS FRAMEWORKS

Administration, Center for Devices and Radiological Health, Office of Device Evaluation, 2016.

Carayon, P., Wetterneck, T. B., Rivera-Rodriguez, A. J., Hundt, A. S., Hoonakker, P., Holden, R., and Gurses, A. P. (2014b). Human factors systems approach to healthcare quality and patient safety. *Appl Ergono*, 45(1), 14–25. DOI: 10.1016/j.apergo.2013.04.023. 9, 10

Chandrasekaran, B. (1990). Design problem solving: A task analysis. *AI Mag* 11.4, 59.

Chipman, S. F., Schraagen, J. M., and Shalin, V. L. (2000). Introduction to cognitive task analysis. *Cognit, Task Anal*, 3–23.

Clark, R. (2014). Cognitive task analysis for expert-based instruction in healthcare. In *Handbook of Research on Educational Communications and Technology*. Springer ,New York, pp. 541–551. DOI: 10.1007/978-1-4614-3185-5_42.

Crandall, B., Klein,G. A., and Hoffman, R. R. (2006). Trends and themes in the development of cognitive task analysis: The rise of modern cognitive psychology. In *Working minds: A Practitioner's Guide to Cognitive Task Analysis*. Mit Press, Cambridge, PA, Ch. 9.

Diller, T., Helmrich, G., Dunning, S., Cox, S., Buchanan, A., and Shappell, S. (2013). The human factors analysis classification system (HFACS) applied to health care. *AJMQ*. DOI: 10.1177/1062860613491623.

Donabedian, A. (1978). The quality of medical care. *Science*, 200(4344), 856–864. DOI: 10.1126/science.417400. 10

Flin, R., Winter, J., and Cakil Sarac, M. R. (2009). Human factors in patient safety: review of topics and tools. *World Health*, 2.

Grudin, J., Pruitt, J., and Adlin, T. (2006). *The Persona Lifecycle: Keeping People in Mind throughout Product Design*. Morgan Kaufmann Publishers Inc., San Francisco, CA.

Jonassen, D. H., Hannum, W. H., and Tessmer, M. (1989) *Handbook of Task Analysis Procedures.* Praeger Publishers.

Karsh, B. T., Weinger, M. B., Abbott, P. A., and Wears, R. L. (2010). Health information technology: fallacies and sober realities. *J AMIA*, 17(6), 617–623. DOI: 10.1136/jamia.2010.005637.

Karsh, B. T., Holden, R. J., Alper, S. J., and Or, C. K. L. (2006). A human factors engineering paradigm for patient safety: designing to support the performance of the healthcare professional. *Qual Safe Health Care*, 15(Suppl 1), i59–i65. DOI: 10.1136/qshc.2005.015974.

Lane, R., Stanton, N. A., and Harrison, D. (2006). Applying hierarchical task analysis to medication administration errors. *Appl Ergono* 37.5, 669–679. DOI: 10.1016/j.apergo.2005.08.001.

Mulder, S. and Yaar, Z. (2007). The user is always right: A practical guide to creating and using personas for the Web. *Interactions-New York,* 14(4), 52–52. 44

Neerincx, M. A. and Griffioen, E. (1996). Cognitive task analysis: harmonizing tasks to human capacities. *Ergonomics* 39.4, 543–561. DOI: 10.1080/00140139608964480.

Nemani, A., Sankaranarayanan, G., Olasky, J. S., Adra, S., Roberts, K. E, Panait, L., Schwaitzberg, S. D., Jones, D. B., and De, S. (2014). A comparison of NOTES transvaginal and laparoscopic cholecystectomy procedures based upon task analysis. *Surg Endosc* 28.8, 2443–2451. DOI: 10.1007/s00464-014-3495-9.

Parush, A., Campbell, C., Hunter, A., Ma, C., Calder, L., Worthington, J., Abbott, C., and Frank, J.R. (2011a). *Situational Awareness and Patient Safety.* Canada: The Royal College of Physicians and Surgeons of Canada, Ottawa, ON. 13, 15

Phipps, D., Meakin, G. H., Beatty, P. C., Nsoedo, C., and Parker, D. (2008). Human factors in anaesthetic practice: insights from a task analysis. *Br J Anaesth,* 100.3, 333–343. DOI: 10.1093/bja/aem392.

Pruitt, J. and Adlin, T. (2010). The Persona Lifecycle: Keeping People in Mind throughout Product Design. Morgan Kaufmann. 44

Salvendy, G. (2012). Task analysis – Why, what and how. In *Handbook of Human Factors and Ergonomics.* John Wiley and Sons, Ch. 13 – DOI: 10.1002/9781118131350.

Shachak, A., Hadas-Dayagi, M., Ziv, A., and Reis, S. (2009). Primary care physicians' use of an electronic medical record system: a cognitive task analysis. *J Gen Int Med,* 24.3, 341–348. DOI: 10.1007/s11606-008-0892-6.

Shappell, S. A. and Wiegmann, D. A. (2012). *A Human Error Approach to Aviation Accident Analysis: The Human Factors Analysis and Classification System.* Ashgate Publishing, Ltd.

Sullivan, M. E., Yates, K. A., Inaba, K., and Clark, R. E. (2014). The use of cognitive task analysis to reveal the instructional limitations of experts in the teaching of procedural skills. *Acad Med*, 89.5, 811–816. DOI: 10.1097/ACM.0000000000000224.

Teixeira, L., Ferreira, C., and Santos, B. S. (2007). Using task analysis to improve the requirements elicitation in health information system. In *Engineering in Medicine and Biology Society*, 2007. 29th Annual International Conference of the IEEE (pp. 3669–3672). DOI: 10.1109/iembs.2007.4353127.

U.S. FDA (2016). Applying human factors and usability engineering to medical devices. U.S. Department of Health and Human Services, Food and Drug Task Analysis and Cognitive Task Analysis.

Vicente, K. J. (1999). *Cognitive Work Analysis: Toward Safe, Productive, and Healthy Computer-based Work*. CRC Press, Boca Raton, FL. 46

Vincent, C., Taylor-Adams, S., and Stanhope, N. (1998). Framework for analyzing risk and safety in clinical medicine. *BMJ*, 316(7138), 1154–1157. DOI: 10.1136/bmj.316.7138.1154. 11

Yu, D., Minter, R. M., Armstrong, T. J., Frischknecht, A. C., Green, C., and Kasten, S. J. (2014). Identification of technique variations among microvascular surgeons and cases using hierarchical task analysis. *Ergonomics*, 57.2 (2014), 219–235. DOI: 10.1080/00140139.2014.884244.

PHYSICAL ENVIRONMENT

Bitner, M. J. (1992). Servicescapes: the impact of physical surroundings on customers and employees. *J Market*, 57–71. DOI: 10.2307/1252042.

Das, B. and Sengupta, A. K.(1996). Industrial workstation design: a systematic ergonomics approach. *Appl Ergon* 27.3, 157–163.

Dascalaki, E. G., Gaglis, A. G., Balaras, C. A., and Lagoudi, A. (2009). Indoor environmental quality in Hellenic hospital operating rooms. *Energy Build*, 41.5, 551–560. DOI: 10.1016/j.enbuild.2008.11.023.

Dijkstra, K., Pieterse, M., and Pruyn, A. (2006). Physical environmental stimuli that turn healthcare facilities into healing environments through psychologically mediated effects: systematic review. *J Adv Nurs*, 56.2, 166–181. DOI: 10.1111/j.1365-2648.2006.03990.x.

Elias, G. A. and S. J. Calil. (2014). Evaluation of Physical Environment Parameters in Healthcare. *XIII Mediterranean Conference on Medical and Biological Engineering and Computing 2013*. Springer International Publishing. DOI: 10.1007/978-3-319-00846-2_292.

Fottler, M. D., Ford, R. C., Roberts, V., and Ford, E. W. (2000). Creating a healing environment: the importance of the service setting in the new consumer-oriented healthcare system. *J Healthcare Mgmt*, 45(2):91–106.

Hickam, David H., Severance, S., Feldstein, A., Ray, L., Gorman, P., Schuldheis, S., Hersh, W. R., Krages, K. P., and Helfand, M. (2003). The effect of health care working conditions on patient safety: Summary. *AHRQ Evidence Report Summaries*.

ISO 9241-11 (1998). *Ergonomic Requirements for Office Work with Visual Display Terminals (VDTs) -- Part 11: Guidance on Usability*. International Organization for Standardization. 91

ISO 7250 (1998). *Basic Human Body Measurements for Technological Design*. International Organization for Standardization.

Koppel, R., Wetterneck, R., Telles, J. L., and Karsh, B-T. (2008). Workarounds to barcode medication administration systems: their occurrences, causes, and threats to patient safety. *J AMIA*, 15.4, 408–423. DOI: 10.1197/jamia.M2616.

Sandberg, W. S., Daily, B., Egan, M., Stahl, J. E., Goldman, J. M., Wiklund, R. A., and Rattner, D. (2005). Deliberate perioperative systems design improves operating room throughput. *Anesthesiology*, 103.2, 406.

Tilley, A. (1993). *The Measure of Man and Woman: Human Factors in Design*. Wiley.

Topf, M. (2000). Hospital noise pollution: an environmental stress model to guide research and clinical interventions. *J Adv Nurs*, 31.3, 520–528. DOI: 10.1046/j.1365-2648.2000.01307.x.

HUMAN ENVIRONMENT

Aiken, L. H. and Patrician, P. A. (2000). Measuring organizational traits of hospitals: the revised nursing work index. *Nurs Res*, 49.3, 146–153. DOI: 10.1097/00006199-200005000-00006.

Baker, D. P., Day, R., and Salas, E. (2006). Teamwork as an essential component of high-reliability organizations. *Health Serv, Res* 41.4p2, 1576–1598. DOI: 10.1111/j.1475-6773.2006.00566.x.

Cardoen, B., Demeulemeester, E., and Beliën, J. (2010). Operating room planning and scheduling: A literature review. *Eur J Op Res*, 201.3, 921–932. DOI: 10.1016/j.ejor.2009.04.011.

Chisholm, C. D., Collison, E. K., Nelson, D. R., and Cordell, W. H. (2000) Emergency department workplace interruptions are emergency physicians "Interrupt-driven" and "Multitasking." *Acad Emerg Med*, 7.11, 1239–1243. DOI: 10.1111/j.1553-2712.2000.tb00469.x.

Chisholm, C. D., Dornfeld, A. M., Nelson, D. R., and Cordell, W. H. (2001). Work interrupted: a comparison of workplace interruptions in emergency departments and primary care offices. *Ann Emer Med*, 38.2, 146–151. DOI: 10.1067/mem.2001.115440.

Duffy, M. C., Azevedo, R., Sun, N-Z., Griscom, S. E., Stead, V., Crelinsten, L., Wiseman, J., Maniatis, T., and Lachapelle, K. (2014). Team regulation in a simulated medical emergency: An in-depth analysis of cognitive, metacognitive, and affective processes. *Instruct Sci*, 1–26. DOI: 10.1007/s11251-014-9333-6.

Grundgeiger, T. and Sanderson, P. (2009). Interruptions in healthcare: theoretical views. *Int J Med Inf*, 78.5, 293–307. DOI: 10.1016/j.ijmedinf.2008.10.001.

Hall, P. and Weaver, L. (2001). Interdisciplinary education and teamwork: a long and winding road. *Med Ed*, 35.9, 867–875. DOI: 10.1046/j.1365-2923.2001.00919.x.

Harrison, M. I., Koppel, R., and Bar-Lev, S.(2007). Unintended consequences of information technologies in health care—an interactive sociotechnical analysis. *J AMIA*, 14.5, 542–549. DOI: 10.1197/jamia.M2384.

Helmreich, R. L. and Merritt, A. R. L. (2001). *Culture at Work in Aviation and Medicine: National, Organizational and Professional Influences*. Routledge.

Hoff, T., Jameson, L., Hannan, E., and Flink, E. (2004). A review of the literature examining linkages between organizational factors, medical errors, and patient safety. *Med Care Res Rev* 61.1:3–37. DOI: 10.1177/1077558703257171.

Kalisch, B. J. and Aebersold, M. (2010). Interruptions and multitasking in nursing care. *Joint Comm J Patient Safety*, 36.3, 126–132. DOI: 10.1016/S1553-7250(10)36021-1.

Kantowitz, B. H. and Sorkin, R. D. (1983). *Human Factors: Understanding People-system Relationships*. John Wiley and Sons, Inc.

Lawton, R. and D. Parker. (2002). Barriers to incident reporting in a healthcare system. *Qual Saf Heath Care*, 11.1, 15–18. DOI: https://doi.org/10.1136/qhc.11.1.15.

Leonard, M., Graham, S., and Bonacum, D. (2004). The human factor: the critical importance of effective teamwork and communication in providing safe care. *Qual Saf Heath Care* 13.suppl 1: i85–i90. DOI: 10.1136/qshc.2004.010033.

Lingard, L., Espin, S., Shyte, S., Regehr, G., Baker G. R., Reznick, R., Bohnen, J., Orser, B., Doran, D., and Grober, E. (2004). Communication failures in the operating room: an observational classification of recurrent types and effects. *Qual Saf Heath Care*, 13.5, 330–334. DOI: 10.1136/qshc.2003.008425.

Manser, T. (2009). Teamwork and patient safety in dynamic domains of healthcare: a review of the literature. *Acta Anaesth Scand,* 53.2, 143–151. DOI: 10.1111/j.1399-6576.2008.01717.x.

Mills, P., Neily, J., and Dunn, E. (2008). Teamwork and communication in surgical teams: implications for patient safety. *JACS,* 206.1, 107–112. DOI: 10.1016/j.jamcollsurg.2007.06.281.

Morey, J. C., Simon, R., Jay, G. D., Wears, R. L., Salisbury, M., Dukes, K. A., and Berns, S. D. (2002). Error reduction and performance improvement in the emergency department through formal teamwork training: evaluation results of the MedTeams project. *Health Svc Res,* 37.6, 1553–1581. DOI: 10.1111/1475-6773.01104.

Neal, A., Griffin, M. A., and Hart, P. M. (2000). The impact of organizational climate on safety climate and individual behavior. *Safety Sci,* 34.1, 99–109. DOI: 10.1016/S0925-7535(00)00008-4.

Nieva, V. F. and J. Sorra. (2003). Safety culture assessment: a tool for improving patient safety in healthcare organizations. *Qual Saf Health Care,* 12 (suppl 2), ii17–ii23. DOI: 10.1136/qhc.12.suppl_2.ii17.

Rivera-Rodriguez, A. J. and Karsh, B. T. (2010). Interruptions and distractions in healthcare: review and reappraisal. *Qual Saf Health Care,* qshc-2009. DOI: 10.1136/qshc.2009.033282.

Salas, E., Rosen, M. A., and King, H.. (2007). Managing teams managing crises: principles of teamwork to improve patient safety in the emergency room and beyond. *Theoret Iss Ergono Sci,* 8.5, 381–394. DOI: 10.1080/14639220701317764.

Salas, E., Dickinson, T.L., Converse, S.A., and Tannenbaum, S.I. (1992). Toward an understanding of team performance and training. In R.W. Swezey and E. Salas (Eds.), *Teams: Their Training and Performance* (pp. 3–29). Ablex , Norwood, NJ.

TECHNOLOGY AND USABILITY

Ash, J. S., Berg, M., and Coiera, E. (2004b) Some unintended consequences of information technology in health care: the nature of patient care information system-related errors. *JAMIA,* 11.2, 104–112. DOI: 10.1197/jamia.M1471.

Fairbanks, R. J. and Caplan, S. (2004). Poor interface design and lack of usability testing facilitate medical error. *Joint Comm J Qual Pat Saf,* 30(10), 579-84. DOI: 10.1016/S1549-3741(04)30068-7.

Ginsburg, G. (2005). Human factors engineering: A tool for medical device evaluation in hospital procurement decision-making. *J Biomed Info,* 38.3, 213–219. DOI: 10.1016/j.jbi.2004.11.008.

Jaspers, M. WM. (2009). A comparison of usability methods for testing interactive health technologies: methodological aspects and empirical evidence. *Int J Med Inf,* 78.5, 340–353. DOI: 10.1016/j.ijmedinf.2008.10.002.

Kushniruk, A. W., Triola, M. M., Borycki, E. M., Stein, R., and Kannry, J. L. (2005). Technology induced error and usability: the relationship between usability problems and prescription errors when using a handheld application. *Int J Med Inf,* 74.7, 519–526. DOI: 10.1016/j.ijmedinf.2005.01.003.

Nielsen, J. (1993). *Usability Engineering.* Elsevier.

Zhang, J., Johnson, T. R., Patel, V. L., Paige, D. L., and Kubose, T. (2003). Using usability heuristics to evaluate patient safety of medical devices. *J Biomed Inf,* 36.1, 23–30. DOI: 10.1016/S1532-0464(03)00060-1.

WORKLOAD, EFFORT, AND RESOURCE DEMAND

Carayon, P. and Gürses, A. P. (2005). A human factors engineering conceptual framework of nursing workload and patient safety in intensive care units. *Int Crit Care Nurs,* 21.5, 284–301. DOI: 10.1016/j.iccn.2004.12.003.

Halford, G. S., Baker, R., McCredden, J. E., and Bain, J. D. (2005) How many variables can humans process? *Psych Sci,* 16.1, 70–76. DOI: 10.1111/j.0956-7976.2005.00782.x.

Halford, G. S., Wilson, W. H., and Phillips, S. (1998) Processing capacity defined by relational complexity: Implications for comparative, developmental, and cognitive psychology. *Behav Brain Sci,* 21.06, 803–831. DOI: 10.1017/S0140525X98001769.

Matthews, G. and Davies, D. R. (2001). Individual differences in energetic arousal and sustained attention: A dual-task study. *Person Indiv Diff,* 31.4, 575–589. DOI: 10.1016/S0191-8869(00)00162-8.

Young, M. S. and Stanton, N. A. (2002). Malleable attentional resources theory: a new explanation for the effects of mental underload on performance. *Hum Fact: J Hum Fact Erg Soc,* 44.3, 365–375. DOI: 10.1518/0018720024497709.

HUMAN CAPABILITIES, LIMITATIONS, AND EMERGENT FACTORS

Ajzen, I. and Fishbein, M. (1977). Attitude-behavior relations: A theoretical analysis and review of empirical research. *Psych Bull,* 84(5), 888. DOI: 10.1037/0033-2909.84.5.888. 80

Dekker, S. (2011). Cognitive factors in healthcare work. In S. Dekker, Ed, *Patient Safety: A Human Factors Approach*. CRC Press, Boca Raton, FL, pp. 65–81, DOI: 10.1201/b10942-4.

Donchin, Y., Gopher, D., Olin, M., Badihi, Y., Biesky, M., Sprung, C. L., Pizov, R., and Cotev, S. (1995) A look into the nature and causes of human errors in the intensive care unit. *Crit Care Med*, 23.2, 294–300. DOI: 10.1097/00003246-199502000-00015.

Drews, F.A. (2012). Human error in healthcare. In P. Carayon, Ed.,, *Handbook of Human Factors and Ergonomics in Healthcare and Patient Safety*, CRC Press, Boca Raton, FL, Ch. 15, pp. 323–340.

Hart, S. G. and Staveland, L. E. (1988). Development of a multi-dimensional workload rating scale: Results of empirical and theoretical research. In P. A. Hancock and N. Meshkati (Eds.), *Human Mental Workload*, Elsevier, Amsterdam, pp. 139–183. 108

Holden, R. J. (2011). Cognitive performance-altering effects of electronic medical records: an application of the human factors paradigm for patient safety. *Cog Tech Work*, 13.1, 11–29.v. DOI: 10.1007/s10111-010-0141-8.

Horsky, J., Zhang, J., and Patel, V. L. (2005).To err is not entirely human: complex technology and user cognition. *J Bio Info*, 38., 264–266. DOI: 10.1016/j.jbi.2005.05.002.

Hunt, E. and D. Waller (1999). Orientation and wayfinding: A review, ONR technical report N00014-96-0380. Office of Naval Research, Arlington, VA.

Laxmisan, A., Hakimzada, F., Sayan, O. R., Green, R. A., Zhang, J., and Patel, V. L. (2007). The multitasking clinician: decision-making and cognitive demand during and after team handoffs in emergency care. *Int J Med Info*, 76.11, 801–811. DOI: 10.1016/j.ijmedinf.2006.09.019.

LeBlanc, V. R., McConnell, M. M., and Monteiro, S. D. (2014). Predictable chaos: a review of the effects of emotions on attention, memory and decision making. *Adv Health Sci Ed*, 1–18. DOI: 10.1007/s10459-014-9516-6.

Mesulam, M-M. (1990). Large-scale neurocognitive networks and distributed processing for attention, language, and memory. *Ann Neur*, 28, 597–613. DOI: 10.1002/ana.410280502.

Montgomery, V. L. (2007). Effect of fatigue, workload, and environment on patient safety in the pediatric intensive care unit. *Ped Crit Care Med*, 8.2, S11–S16. DOI: 10.1097/01. PCC.0000257735.49562.8F.

Parush, A., Campbell, C., Hunter, A., Ma, C., Calder, L., Worthington, J., Abbott, C., and Frank, J.R. (2011b). *Situational Awareness and Patient Safety*. The Royal College of Physicians and Surgeons of Canada, Ottawa, ON, Canada.

Peters, G.A. and Peters, B.J. (2008). *Medical Error and Patient Safety: Human Factors in Medicine.* : Human Factors, CRC Press, Boca Raton, FL, Ch. 6 pp. 95–120.

Reason, J. (1995). Understanding adverse events: human factors. *Qual Health Care,* 4.2, 80–89. DOI: 10.1136/qshc.4.2.80.

Sexton, J. B., Thomas, E. J., and Helmreich, R. L. (2000). Error, stress, and teamwork in medicine and aviation: cross sectional surveys. *BMJ,* 320.7237, 745–749. DOI: 10.1136/bmj.320.7237.745.

World Health Organization (2009). *Human Factors in Patient Safety: Review for Methods and Measures.* 12

Yerkes, R. M. and Dodson, J. D. (1908). The relationship of strength of stimulus to rapidity of habit formation. *J Comp Neuro Psych,* 18, 459–482. DOI: 10.1002/cne.920180503. 111

OUTCOMES

Elder, N. C. and Dovey, S. M. (2002). Classification of medical errors and preventable adverse events in primary care: a synthesis of the literature. *J Fam Prac,* 51(11), 927–932. 123

Mishra, A., Catchpole, K., and McCulloch, P. (2009). The Oxford NOTECHS System: reliability and validity of a tool for measuring teamwork behaviour in the operating theatre. *Qual Safe Health Care,* 18.2, 104–108. DOI: 10.1136/qshc.2007.024760.

Norman, D. A. (1988). *The Psychology of Everyday Things.* Basic Books.

Øvretveit, J. (2011). Understanding the conditions for improvement: research to discover which context influences affect improvement success. *BMJ Qual Saf,* 20 (Suppl 1) i18–i23. DOI: 10.1136/bmjqs.2010.045955.

Reason, J. (1990). *Human Error.* Cambridge University Press. DOI: 10.1017/cbo9781139062367. 122

Reason, J. (2000). Human error: Models and management. *BMJ,* 320(7237), 768–77. DOI: 10.1136/bmj.320.7237.768. 122

Schouten, L. M. T., Hulscher, M. E. J. L., van Everdingen, J. J. E., Huijsman, R., and Grol, R. P. T. M. (2008). Evidence for the impact of quality improvement collaboratives: systematic review. *BMJ,* 336.7659, 1491–1494. DOI: 10.1136/bmj.39570.749884.BE.

TRADEOFF ANALYSIS

Hagen, C. W. (1967). *Techniques for the Allocation and Evaluation of Human Resources.* Martin Marietta Corporation (Report OR 8735), Orlando, Florida. 136

Meister, D. (1985). *Behavioral Analysis and Measurement Methods*. Wiley-Interscience. 136

Pugh, S. (1991). *Total Design: Integrated Methods for Successful Product Engineering* Addison-Wesley, Wokingham, MA, (p. 278). 136

Otto, K. N. and Wood, K. (2003). *Product Design: Techniques in Reverse Engineering and New Product Development*. Prentice Hall. 136

Author Biographies

Avi Parush is an associate professor at the Industrial Engineering and Management Faculty, The Israel Institute of Technology, and an emeritus professor from the psychology department of Carleton University, Ottawa, Canada. With over 30 years of practice and research in human factors, HCI, and usability, Avi devoted his career to influencing the design of workplaces and the tools people use to make their lives easier, safer, and more beneficial. His current research focuses on teamwork in complex and critical situations in healthcare, first-response and other domains, and simulation-based training with a focus on team training. He is the emeritus founding editor-in-chief of the *Journal of Usability Studies*, and is currently on the editorial board of the *Human Factors Journal*.

Debi Parush analyzes and designs knowledge-sharing systems to fill gaps and constantly improve. Debi has practiced listening and questioning for insights, and synthesizing and designing training and other performance support solutions in Israeli and Canadian high tech industries as well as in public service. Debi has a Master Degree's in Educational Technology—bridging the gap between where we are and where we need to be—and a Graduate Certificate in Conflict Resolution—getting there together sometimes requires facilitation. For many years, Debi has managed a consulting firm specializing in human factors, human-computer interaction design, developing training and marketing materials, and editorial services.

Dr. Roy Ilan practices Internal Medicine and Critical Care Medicine, and provides supervision to students and residents at various training levels in both specialties at Kingston General Hospital, a tertiary care teaching center affiliated with Queen's University. He co-chairs the patient safety and quality improvement (PS&QI) committees for the Departments of Medicine and Critical Care Medicine. His research is focused on patient safety and quality improvement, and he has implemented system-level interventions to address various issues. Dr. Ilan has taught extensively in the area of PS&QI at different organizations, including the Canadian Patient Safety Institute (CPSI), the Royal College of Physicians and Surgeons of Canada, and Queen's University.

Printed in the United States
by Baker & Taylor Publisher Services